Daniel Höhne-Mönch

Steady-state emission of blazars at very high energies

Daniel Höhne-Mönch

Steady-state emission of blazars at very high energies

Analysis and interpretation of observations performed with the MAGIC telescope

Südwestdeutscher Verlag für Hochschulschriften

Imprint
Any brand names and product names mentioned in this book are subject to trademark, brand or patent protection and are trademarks or registered trademarks of their respective holders. The use of brand names, product names, common names, trade names, product descriptions etc. even without a particular marking in this work is in no way to be construed to mean that such names may be regarded as unrestricted in respect of trademark and brand protection legislation and could thus be used by anyone.

Publisher:
Südwestdeutscher Verlag für Hochschulschriften
is a trademark of
Dodo Books Indian Ocean Ltd., member of the OmniScriptum S.R.L Publishing group
str. A.Russo 15, of. 61, Chisinau-2068, Republic of Moldova Europe
Printed at: see last page
ISBN: 978-3-8381-2505-3

Zugl. / Approved by: Würzburg, Julius-Maximilians-Universität, Dissertation, 2010

Copyright © Daniel Höhne-Mönch
Copyright © 2011 Dodo Books Indian Ocean Ltd., member of the OmniScriptum S.R.L Publishing group

Contents

Abstract   5

Zusammenfassung   7

Introduction   9

1 Evolution of blazars   11
   1.1 Active galactic nuclei 11
      1.1.1 Empirical classification 11
      1.1.2 Unified scheme for AGN 13
   1.2 Blazars 13
      1.2.1 Emission models 14
      1.2.2 Blazar sequence 16
      1.2.3 BL Lac objects 18
   1.3 Blazar unification by evolution 19

2 The very high energy γ-ray and cosmic ray connection   23
   2.1 Cosmic rays 23
   2.2 Very high energy γ-rays 26
      2.2.1 Production mechanisms 27
      2.2.2 Intergalactic absorption of γ-rays 28
   2.3 Sources 28
      2.3.1 γ-ray bursts 29
      2.3.2 Starburst galaxies 30
      2.3.3 Compact objects 30
      2.3.4 Active galactic nuclei 31
      2.3.5 Diffuse emission 33

Contents

3 Imaging atmospheric Cherenkov technique — 35
3.1 Extensive air showers — 35
3.2 Cherenkov effect — 36
3.3 Imaging technique — 39
3.4 MAGIC telescope — 39
3.4.1 Structure and Reflector — 40
3.4.2 Camera — 41
3.4.3 Data acquisition and trigger system — 41
3.4.4 Observation modes and file types — 42
3.4.5 Monte Carlo simulations — 44

4 Analysis chain — 45
4.1 Signal extraction and calibration — 45
4.1.1 Signal extraction — 46
4.1.2 Calibration — 46
4.1.3 Bad pixel treatment — 47
4.2 Event image reconstruction — 48
4.2.1 Software trigger — 48
4.2.2 Image cleaning — 48
4.2.3 Image parametrisation — 49
4.3 Background rejection — 51
4.3.1 Quality cuts — 52
4.3.2 γ – hadron separation cuts — 52
4.4 Spectrum — 53
4.4.1 Energy estimation — 53
4.4.2 Effective collection area — 54
4.4.3 Effective observation time — 54
4.4.4 Energy spectrum — 54
4.5 Lightcurves — 55
4.6 Observations of the Crab Nebula — 55
4.6.1 Data selection and automatic analysis — 55
4.6.2 Background rejection — 56
4.6.3 Energy spectrum — 56

5 Observations and analysis results — 59
5.1 Search for TeV candidate BL Lac objects — 60

		5.1.1	TeV flux estimation	60

- 5.1.1 TeV flux estimation 60
- 5.1.2 Source catalogues and compilations 61
- 5.1.3 Selection criteria 62
- 5.2 Observation campaign 65
  - 5.2.1 Known sources from the selected sample 66
  - 5.2.2 Tentative redshift measurements 69
- 5.3 Analysis results 70
  - 5.3.1 Results of the analysis chain 70
  - 5.3.2 Upper limit calculation 76
  - 5.3.3 Significance distribution 76
  - 5.3.4 Source stacking 77
  - 5.3.5 Crosscheck analysis 80

6 Steady state emission of blazars   85
- 6.1 Spectral characteristics 85
  - 6.1.1 multi-wavelength data 85
  - 6.1.2 EBL correction 88
  - 6.1.3 Result 89
- 6.2 Comparison with known steady state sources 94
  - 6.2.1 HBLs measured in a low emission state 94
  - 6.2.2 Broad-band spectral indices 95
  - 6.2.3 Spectral energy distribution 96

7 Conclusions and outlook   99

A Data compendium   101

B ϑ^2-distributions   123

C Lightcurves   137

List of figures   149

List of tables   151

Bibliography   153

List of publications   165

Contents

Acknowledgements **173**

Summary

One key scientific program of the MAGIC telescope project is the discovery and detection of blazars. They constitute the most prominent extragalactic source class in the very high energy (VHE) γ-ray regime with 29 out of 34 known objects[1]. Therefore a major part of the available observation time was spent in the last years on high-frequency peaked blazars. The selection criteria were chosen to increase the detection probability. As the X-ray flux is believed to be correlated to the VHE γ-ray flux, only X-ray selected sources with a flux $F_X > 2\,\mu$Jy at 1 keV were considered. To avoid strong attenuation of the γ-rays in the extragalactic infrared background, the redshift was restricted to values between $z < 0.15$ and $z < 0.4$, depending on the declination of the objects. The latter determines the zenith distance during culmination which should not exceed 30° (for $z < 0.4$) and 45° (for $z < 0.15$), respectively.

Between August 2005 and April 2009, a sample of 24 X-ray selected high-frequency peaked blazars has been observed with the MAGIC telescope. Three of them were detected including 1ES 1218+304 being the first high-frequency peaked BL Lacertae object (HBL) to be discovered with MAGIC in VHE γ-rays. One previously detected object was not confirmed as VHE emitter in this campaign by MAGIC. A set of 20 blazars previously not detected will be treated more closely in this work. In this campaign, during almost four years ∼ 450 hrs or ∼ 22 % of the available observation time for extragalactic objects were dedicated to investigate the baseline emission of blazars and their broad-band spectral properties in this emission state. For the sample of 20 objects in a redshift range of $0.018 < z < 0.361$ integral flux upper limits in the VHE range on the 99.7 % confidence level (corresponding to 3 standard deviations) were calculated resulting in values between 2.9 % and 14.7 % of the integral flux of the Crab Nebula.

As the distribution of significances of the individual objects shows a clear shift to positive values, a stacking method was applied to the sample. For the whole set of 20 objects, an excess of γ-rays was found with a significance of 4.5 standard deviations in 349.5 hours of effective

---
[1] As of April 2010

Summary

exposure time. For the first time a signal stacking in the VHE regime turned out to be successful. The measured integral flux from the cumulative signal corresponds to 1.4 % of the Crab Nebula flux above 150 GeV with a spectral index $\alpha = -3.15 \pm 0.57$. None of the objects showed any significant variability during the observation time and therefore the detected signal can be interpreted as the baseline emission of these objects.

For the individual objects lower limits on the broad-band spectral indices $\alpha_{X-\gamma}$ between the X-ray range at 1 keV and the VHE γ-ray regime at 200 GeV were calculated. The majority of objects show a spectral behaviour as expected from the source class of HBLs: The energy output in the VHE regime is in general lower than in X-rays. For the stacked blazar sample the broad-band spectral index was calculated to $\alpha_{X-\gamma} = 1.09$, confirming the result found for the individual objects. Another evidence for the revelation of the baseline emission is the broad-band spectral energy distribution (SED) comprising archival as well as contemporaneous multi-wavelength data from the radio to the VHE band. The SEDs of known VHE γ-ray sources in low flux states matches well the SED of the stacked blazar sample.

Zusammenfassung

Eines der wissenschaftlichen Schlüsselprogramme des MAGIC Projektes ist die Entdeckung und Detektion von Blazaren. Diese stellen mit 29 von 34 bekannten Objekten die prominenteste extragalaktische Quellklasse im Bereich der sehr hochenergetischen (engl. very high energy, VHE) γ-Strahlung dar. Deshalb wurde in den letzten Jahren ein Großteil der verfügbaren Beobachtungszeit sogenannten Blazaren mit hochfrequenten Peaks (engl. high-frequen-cy peaked) gewidmet. Die Auswahlkriterien dafür wurden entsprechend ge-wählt, um die Detektionswahrscheinlichkeit zu erhöhen. Da man glaubt, dass der Röntgenfluss mit dem VHE γ-Fluss korreliert, wurden nur röntgenselek-tierte Quellen mit einem Fluss $F_X > 2\,\mu$Jy bei 1 keV betrachtet. Um eine starke Abschwächung der γ-Strahlung innerhalb des extragalaktischen Infrarot-Hintergrundes zu vermeiden, wurde die Rotverschiebung auf Werte zwischen $z < 0{,}15$ und $z < 0{,}4$ begrenzt, abhängig von der Deklination der Objekte. Diese bestimmt den Zenitdistanz während der Kulmination, der $30°$ (für $z < 0{,}15$) bzw. $45°$ (für $z < 0{,}4$) nicht übersteigen sollte.

Zwischen August 2005 und April 2009 wurde ein Sample aus 24 röntgenselek-tierten high-frequency peaked Blazaren mit dem MAGIC Teleskop beobachtet. Drei davon wurden detektiert, einschließlich 1ES 1218+304, der erste HBL (engl. von high-frequency peaked BL Lacertae object), der mit MAGIC im VHE γ-Bereich entdeckt wurde. Ein früher entdecktes Objekt konnte in dieser Kampagne nicht von MAGIC als VHE Emitter bestätigt werden. Ein Set aus 20 im Vorfeld nicht detektierten Blazaren wird in dieser Arbeit genauer betrachtet. Während fast vier Jahren wurden in dieser Kampagne ~ 450 h oder $\sim 22\,\%$ der verfügbaren Beobachtungszeit für extragalaktische Objekte der Untersuchung der Grundzustandsemission von Blazaren und deren breitband-spektralen Eigenschaften in diesem Zustand gewidmet. Für das Sample aus 20 Objekten in einem Rotverschiebungsbereich $0.018 < z < 0.361$ wurden integrale Flussobergrenzen im VHE Bereich auf Basis eines $99{,}7\,\%$ Konfidenzlevels (entsprechend 3 Standardabweichungen) berechnet. Damit liegen die Obergrenzen zwischen $2{,}9\,\%$ und $14{,}7\,\%$ des integralen Flusses des Krebsnebels.

Zusammenfassung

Da die Verteilung der Signifikanzen der einzelnen Objekte eine klare positive Verschiebung aufweist, wurde eine Stacking-Methode auf das Sample angewandt. Für das gesamte Set aus 20 Objekten konnte ein γ-Strahlungsexzess mit einer Signifikanz von 4,5 Standardabweichungen bei einer effektiven Be-obachtungszeit von 349,5 h gefunden werden. Zum ersten Mal war ein Signal-Stacking im VHE Bereich erfolgreich. Der gemessene integrale Fluss des kumulativen Signals entspricht 1,4 % des Flusses des Krebsnebels oberhalb einer Energie von 150 GeV mit einem Spektralindex $\alpha = -3{,}15 \pm 0{,}57$. Keines der Objekte zeigte Anzeichen für Variabilität während der Beobachtungszeit und daher kann das detektierte Signal als die Grundzustandsemission dieser Objekte angesehen werden.

Für die einzelnen Objekte wurden untere Grenzen für die Breitband-Spektral-indizes $\alpha_{X-\gamma}$ zwischen dem Röntgenbereich bei 1 keV und dem VHE γ-Bereich bei 200 GeV berechnet. Die Mehrheit der Objekte zeigt ein spektrales Verhalten, wie es für die Klasse der HBLs erwartet wird: Der Energieausstoß im VHE γ-Bereich is im allgemeinen niedriger als im Röntgenbereich. Für das mit dem Stacking betrachtete Blazar-Sample wurde der Breitband-Spektralindex zu $\alpha_{X-\gamma} = 1{,}09$ berechnet, was die Ergebnisse für die einzelnen Objekte bestätigt. Ein weiterer Hinweis für die Aufdeckung der Grundzustandsemission is die breitband-spektrale Energieverteilung (engl. spectral energy distribution, SED), die Archiv- wie auch kontemporäre Multiwellenlängendaten vom Radio- bis in den VHE γ-Bereich enthält. Die SEDs bekannter VHE γ-Quellen in niedrigen Flusszuständen stimmt gut mit der SED aus dem Stacking des Blazar-Samples überein.

Introduction

The young field of very high energy (VHE) γ-ray astronomy started in the 1960s with the first detection of air showers by Chudakov et al. (www-01). The VHE γ-ray range above $\sim 100\,\text{GeV}$ is hardly accessible via satellite or balloon borne instruments which paved the way for the atmospheric Cherenkov technique. The first telescope of this kind – Whipple – was constructed in 1968 (www-01) and enabled the discovery of the Crab Nebula and Mkn 421 as the first galactic and extragalactic VHE γ-ray sources, respectively (Weekes et al., 1989; Punch et al., 1992).

Since then, ground-based VHE γ-ray astronomy is very successful in revealing the most violent and powerful phenomena in the universe. In particular the blazar-type BL Lacertae objects, a subclass of active galactic nuclei (AGN), turned out to be rich targets in this energy range. The MAGIC telescope, one of the new generation γ-ray experiments, aims at discovering new objects and source classes as well as investigating the fundamental characteristics of blazars.

Only in the last years, the increased sensitivity of new ground-based VHE instruments such as MAGIC, H.E.S.S. and VERITAS rendered the detection of low or steady emission states of blazars possible. However, many BL Lacertae objects eluded their discovery and remained undetected in the VHE range. Therefore the question arises, if BL Lacertae objects do feature a steady-state emission in general or only appear in active, so-called flaring states.

The following work tries to answer this question by investigating a sample of previously undetected BL Lacertae objects with MAGIC. It is organised as follows:

> In chapter 1 AGN are presented and classified in general and a subclass of them – blazars – is treated in particular. The chapter deals with possible emission models of γ-rays from these objects and the spectral sequence of blazars introduced by Fossati et al. (1998). In addition to this purely spectral unification, the temporal evolution of blazars is considered as basis for a unification.

Introduction

A consequence of the evolutionary sequence of blazars is the emission of both cosmic rays and VHE γ-rays. After a discussion of the cosmic ray spectrum the production and propagation of VHE γ-rays is presented in chapter 2, followed by possible and known sources for both kinds of radiation.

Chapter 3 covers the development of extended air showers initiated by VHE γ-rays and their detection principle. The imaging atmospheric Cherenkov technique makes use of the Cherenkov effect occurring within extended air showers in the atmosphere. Furthermore the MAGIC telescope as a detector of VHE γ-rays is brought into focus.

The determination of the primary particle type hitting the atmosphere and its energy is achieved with a partially automated analysis chain explained in chapter 4. The Cherenkov light signals are extracted from the recorded data and being calibrated. Afterwards, the event images are reconstructed in the MAGIC camera plane and parametrised. The background of non-γ-like events is then suppressed and from the remaining events the energy can be estimated as well as a spectrum calculated. Being one of the strongest and a constant VHE γ-ray source in the sky, the Crab Nebula is analysed as proof of concept and for comparative studies in the following chapters.

Chapter 5 deals with the observation campaign conducted to discover new VHE γ-ray sources and unveil the characteristics of their baseline emission. Promising candidates for this campaign have been selected from a set of catalogues and compilations. The focus lies on the analysis of previously undetected BL Lac objects. For a set of 20 objects integral upper limits are calculated. The baseline emission of the cumulative sample is revealed by applying a stacking method to the data.

The baseline or steady-state emission of BL Lac objects is treated more closely in chapter 6. The spectral characteristics are investigated inferring multi-wavelength data from the radio to the X-ray regime as well as the baseline energy spectrum in the VHE regime, corrected for γ-ray absorption in the extragalactic background light. The multi-wavelength approach allows to calculate broad-band spectral indices for the individual objects as well as for the cumulative sample and to assemble a broad-band spectral energy distribution which can be compared to known VHE steady-state sources.

Chapter 7 summarises the key findings of this work. It gives an outlook on the impact of future experiments and possible tasks implied by this work.

1 Evolution of blazars

Blazars belong to the most extreme objects in the γ-ray sky. The expression 'blazar' is deduced from the term 'blazing quasi-stellar object' and was introduced by Edward Spiegel in 1978 (Angel and Stockman, 1980). Their energy output can amount to $10^{49}\,\mathrm{erg\,s^{-1}}$ assuming isotropic emission. A particular subclass, the high frequency peaked BL Lacertae objects (HBLs), is characterised by emission of γ-rays in the very high energy (VHE) regime above $\sim 100\,\mathrm{GeV}$. Therefore blazars are one of the most important extragalactic source classes to observe with imaging atmospheric Cherenkov telescopes (cf. chapter 3).

As blazars belong to the larger group of AGN, the latter ones will be introduced and classified in the first section of this chapter. Afterwards blazars themselves and different scenarios for their unification will be described in sections 1.2 and 1.3.

1.1 Active galactic nuclei

AGN comprise $\sim 1\%$ of all known galaxies. They are characterised by a bright bulge, the central galactic region, which is brighter compared to the same Hubble type galaxies. The emitted photon spectrum in general shows two components: a nonthermal continuum from radio to X-rays or γ-rays and an optical–X-ray continuum of thermal origin. Morphologically, the nonthermal emission can be associated with the jets, plasma outflows from the active galactic nucleus, whereas the thermal component, the so-called Big Blue Bump, presumably originates from an accretion disk surrounding a supermassive black hole. In addition, emission lines are present due to clouds photonised by the central continuum. Eventually, AGN show variability on all wavelength- and timescales. The different subclasses do not show necessarily all of these characteristics.

1.1.1 Empirical classification

The empirical classification as well as the following description of a unified scheme for radio-loud AGN is made according to Urry and Padovani (1995). AGN can be classified by means

1 Evolution of blazars

of their radio loudness and optical spectra showing broad (type 1), narrow (type 2) or unusual (type 0) emission lines. The radio loudness is defined as the ratio of the flux $F_{5\,\text{GHz}}$ at 5 GHz to the one in the optical B band, F_B. An AGN is called radio-loud in case of $F_{5\,\text{GHz}}/F_B \gtrsim 10$, radio-quiet otherwise. 15 - 20 % of all AGN are radio-loud.

Table 1.1 lists the different AGN types. The upper part describes the radio-quiet, the lower part the radio-loud populations. The structure from left to right is chosen such that the angle of the line of sight to the observer is decreasing, according to the AGN paradigm presented in the next section.

Type 2 AGN have weak continua and show only narrow line emission. The sight on broad emission line regions is obscured by absorbing material due to the edge-on line of sight to the observer. In this class one finds at low luminosities Seyfert 2 galaxies and narrow-emission-line X-ray galaxies (NELG). Candidates at high luminosities might be infrared-luminous quasars. The radio-loud type 2 AGN are narrow-line radio galaxies (NLRG) consisting of two morphologically different types: low-luminosity FR I and high-luminosity FR II galaxies (FR: Fanaroff-Riley type, Fanaroff and Riley (1974)).

radio loudness	type 2 (narrow lines)	type 1 (broad lines)	type 0 (unusual)
radio-quiet	Seyfert 2 NELG IR quasar?	Seyfert 1 QSO	BAL QSO?
radio-loud	FR I FR II	BLRG SSRQ FSRQ	blazars: BL Lac objects (FSRQ)

Table 1.1: AGN taxonomy, taken from Urry and Padovani (1995)

In Type 1 AGN – having bright continua – hot gas with high velocities in the nuclear region produces broad emission lines. In case of radio-quiet AGN these are Seyfert 1 galaxies at low and quasars (QSO, quasi-stellar objects) at higher luminosities. The radio-loud low-luminosity sources are called broad-line radio galaxies (BLRG) and the high-luminosity quasars are either steep spectrum or flat spectrum radio quasars (SSRQ and FSRQ, respectively). The separation is done at a radio spectral index $\alpha_r = 0.5$.

Being the smallest group among AGN, the type 0 shows unusual spectral features. Presumably they are oriented with a very small angle to the line of sight to the observer. Roughly 10 % of radio-quiet AGN are known as broad absorption line quasars (BAL QSO). The radio-loud type 0 population consists of blazars, mainly BL Lacertae objects. There is also a subset of type 1 FSRQs seen under a small angle to the line of sight. BL Lac objects will be treated in more detail in section 1.2.3.

Radio-loud AGN like the FR I/II type are mostly hosted by elliptical galaxies whereas radio-quiet ones like Seyfert galaxies can also be spiral galaxies with unusually bright core regions. Blazar host galaxies are difficult to determine because the core emission outshines the host galaxy by far. The fact, that radio-loud AGN and thus presumably also blazars are mostly found in elliptical galaxies leads to the conclusion that a turbulent evolution of merging events could be the initial spark for the genesis of these violent objects. This topic will be revisited in section 1.3.

1.1.2 Unified scheme for AGN

In the unified scheme of AGN a central supermassive black hole (SMBH) of $\sim 10^8$ solar masses ($M_\odot$) accretes matter which forms an accretion disc around the SMBH. Inside the accretion disc gravitational energy is transformed into thermal radiation. Rapidly moving gas clouds near the SMBH serve as target for atomic excitation or ionisation. They produce broad emission lines and are thus called BLRs. Further away from the SMBH slower gas clouds exhibit narrow line emission (NLRs). The central region is surrounded by a torus of dust which absorbs emission from the vicinity of the SMBH in the equatorial plane. Perpendicular to the accretion disc there are two ultra-relativistic plasma outflows called jets. Due to relativistic beaming, accelerated particles produce strongly collimated VHE γ-ray emission along the jet axis (cf. section 2.2). Figure 1.1 illustrates the AGN paradigm. Based on the line of sight to the observer different AGN populations can be observed. The typical parameters for an AGN with a $10^8 \, M_\odot$ SMBH can be found in table 1.2.

1.2 Blazars

The AGN subclass of blazars includes FSRQs and BL Lac objects (one and 28 sources among them detected in the VHE regime by April 2010, respectively). Most extragalactic VHE γ-ray sources are blazars. Two sources are starburst galaxies and another two sources radio galaxies (including the giant FR-I radio galaxy M87 which could be interpreted as a 'misaligned' blazar).

1 Evolution of blazars

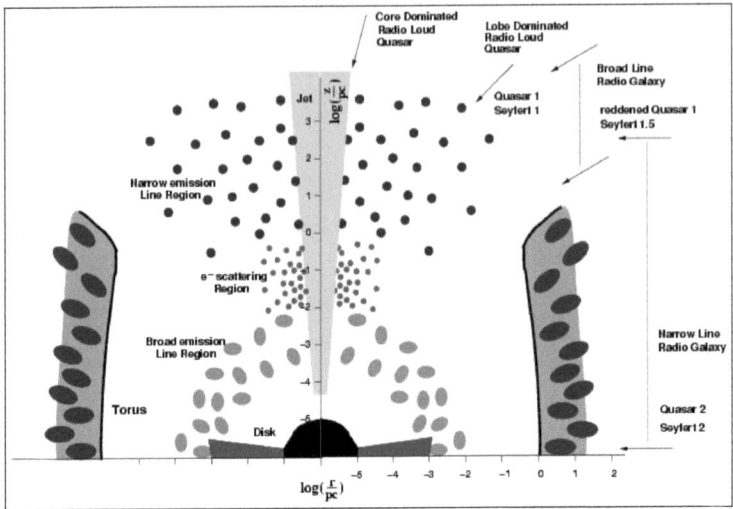

Figure 1.1: The unified AGN scheme. A supermassive black hole in the centre is surrounded by an accretion disc and a dusty torus. Perpendicular to the accretion plane a collimated jet is ejected where particle acceleration up to relativistic energies takes place. Additionally broad and narrow emission line regions complete the scheme. An AGN classification can be done depending on the viewing angle. Taken from Zier and Biermann (2002).

1.2.1 Emission models

The spectral energy distribution (SED) of blazars is characterised by nonthermal continuum emission extending over 20 orders of magnitude. It shows two pronounced humps in a νF_ν vs ν diagram, the first one at IR to X-ray energies, the second one at γ-rays. Figure 1.2 gives an exemplary SED of a blazar.

In order to explain the structure of the SED different emission models can be considered.

component	size in cm
radius of SMBH	$\sim 3 \times 10^{13}$
radius of accretion disc	$\sim 1 - 30 \times 10^{14}$
distance of broad-line region	$\sim 2 - 20 \times 10^{16}$
inner radius of dusty torus inner radius	$\sim 10^{17}$
extension of narrow-line region	extending from 10^{18} to 10^{19}
extension of radio jets	up to 10^{24}

Table 1.2: Typical sizes of the different AGN components for a central black hole with 10^8 M$_\odot$ according to Urry and Padovani (1995).

1.2 Blazars

The low energy peak usually is ascribed to synchrotron radiation from electrons. The high energy peak, however, gives rise to different possible scenarios based on either leptonic or hadronic induced γ-ray emission.

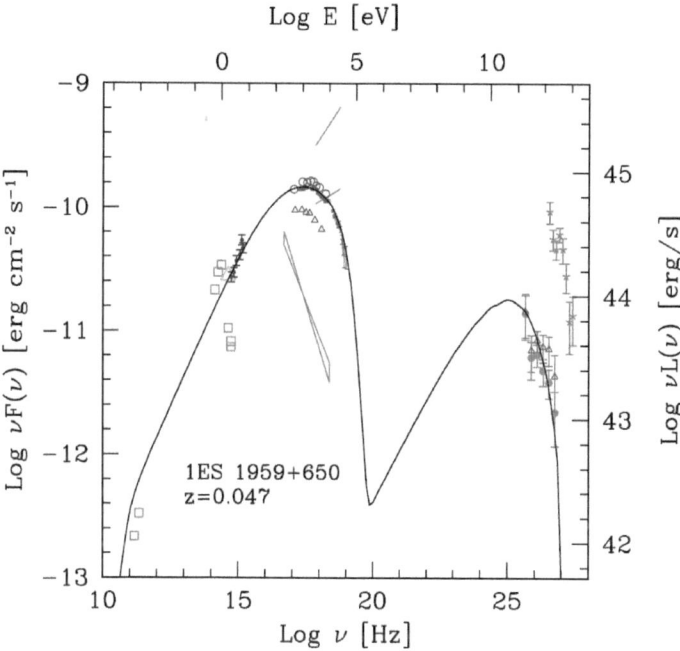

Figure 1.2: Spectral energy distribution of the BL Lac object 1ES 1959+650 as measured during a multi-wavelength campaign in 2006 (Tagliaferri et al., 2008).

Leptonic models

In leptonic models inverse Compton (IC) scattering of lower energy photons (optical to X-rays) leads to VHE γ-ray emission. In self-synchrotron Compton models (SSC) the synchrotron radiation of ultra-relativistic electrons responsible for the low energy peak serves as target photon field for the IC process (Ginzburg and Syrovatskii, 1965; Rees, 1967; Maraschi et al., 1992). Another possibility is external Compton (EC) radiation, for instance emitted by nearby stellar populations or gas clouds, that is upscattered by the electrons into the VHE regime (Dermer et al., 1992; Dermer and Schlickeiser, 1993).

Hadronic models

In hadronic models the jet contains an ultra-relativistic protonic component. These pro-

1 Evolution of blazars

tons can also emit synchrotron radiation, but due to their high energy of $\gtrsim 10^{18}$ eV the emission is in the VHE range (synchrotron proton blazar, SPB, Mannheim, 1993; Mücke and Protheroe, 2001). Another possibility for hadronic interactions are cascades induced by ultra high energy (UHE) protons where subsequent π^0-decay leads to γ-radiation that interacts further with low energy photons in the jet (proton induced cascades, PIC, Mannheim, 1993). The detection of neutrinos produced in hadronic interactions could hint on hadronic acceleration mechanisms in blazar jets. In addition, hadronic models give a natural explanation for the existence of UHE cosmic rays detected on earth.

1.2.2 Blazar sequence

Fossati et al. (1998) investigated the spectral properties of 126 blazars and found a correlation of several characteristics with one key parameter, the radio luminosity at 5 GHz L_r. They computed the SEDs for three complete samples – the Einstein slew survey sample (Elvis et al., 1992), the 1-Jy sample of BL Lacs (Kuehr et al., 1981) and the sample of FSRQs derived from Wall and Peacock (1985) – and for groups of blazars binned in radio luminosity independent of their classification. The result can be seen in figure 1.3: (i) The frequency and luminosity of the first peak correlates with the radio luminosity in the sense that with increasing L_r the frequency $\nu_{\mathrm{peak,sync}}$ increases and the corresponding luminosity decreases. (ii) The peak frequency $\nu_{\mathrm{peak},\gamma}$ of the γ-component correlates with the peak frequency $\nu_{\mathrm{peak,sync}}$ of the lower energy one. That is the ratio $\nu_{\mathrm{peak},\gamma}/\nu_{\mathrm{peak,sync}}$ is consistent with a constant. (iii) Finally the ratio of the luminosity at the peaks, $L_{\mathrm{peak},\gamma}/L_{\mathrm{peak,sync}}$, increases with L_r.

Agreement with the blazar spectral sequence is found by Ghisellini et al. (1998, 2002). They simulated blazar SEDs by means of SSC and EC models finding a strong correlation between the energy of the electrons emitting at the synchrotron peak, γ_{peak}, and the Compton dominance being the ratio L_C/L_{sync} of the luminosities at the Compton and synchrotron peaks, respectively. Additionally the lowest power blazars could be well explained adding a finite injection time of the relativistic electrons in the source region.

Several authors argued against a spectral sequence proposed by Fossati et al. (1998) being at least partially a selection effect. Giommi et al. (2002a) conclude that the correlation between the position of the synchrotron peak and the radio luminosity is weak because they find a comparable fraction of HBLs in each radio luminosity bin in the DXRBS (Deep X-ray Radio Blazars Survey, Perlman et al., 1998; Landt et al., 2001), the Sedentary multi-frequency survey (Giommi et al., 1999) and the NVSS-RASS cross-correlation (NRAO VLA Sky Survey –

ROSAT All Sky Survey, Giommi et al., 2002b). Especially the emergence of HBL-like FSRQs (HFSRQ, Perlman et al., 1998; Padovani et al., 2002, 2003) support these findings. Also Caccianiga and Marchã (2004) find low power blazars outside the spectral blazar sequence in the CLASS Blazar Survey (Cosmic Lens All-Sky Survey, Marchã et al., 2001).

Recently Giommi et al. (2007) and Bassani et al. (2007) reported the existence of two high redshift FSRQs with unusually high synchrotron peak frequencies in the X-ray domain. They concluded that these blazars lie outside the spectral sequence. However, Maraschi et al. (2008a,b) argued that the SEDs of these extraordinary blazars can well be explained within the sequence when assigning the X-ray measurement to the rising edge of the Compton peak. A new perspective on the spectral blazar sequence was driven by Ghisellini and Tavecchio (2008) taking into account more physical parameters for simulating successfully the whole spectrum of different blazar SEDs including low luminosity 'red' quasars and 'blue' quasars with broad emission

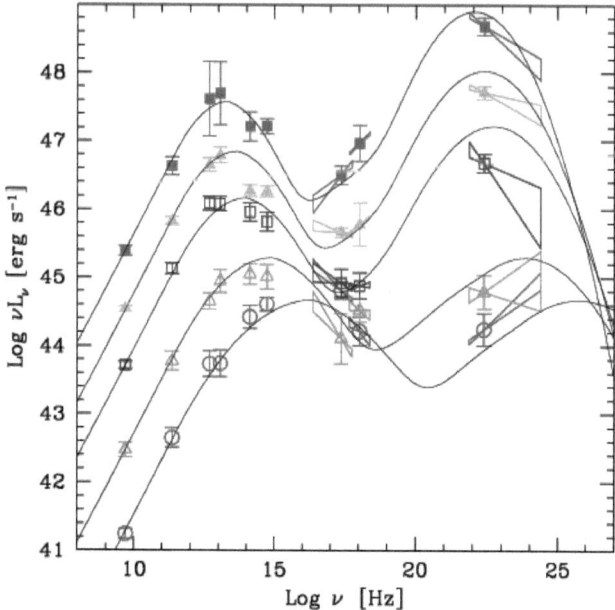

Figure 1.3: Spectral blazar sequence as introduced by Fossati et al. (1998). With increasing radio luminosity the synchrotron and inverse Compton peak frequencies are decreasing whereas their luminosities increase. Taken from Donato et al. (2001).

1 Evolution of blazars

lines but with similar SEDs as low luminosity HBLs. This scenario leads to a spectral blazar sequence dependent of two physical parameters, the mass m and the accretion rate $\dot{m}$ of the central black hole, based on some simplifying assumptions: (i) The kinetic power of the jet is proportional to $\dot{m}$; (ii) Most of the jet dissipation takes place at a distance from the black hole that is proportional to m; (iii) The BLR exists only above a critical value of the disk luminosity; (iv) The radius of the BLR scales with the disk luminosity squared.

A large caveat of the spectral blazar sequence introduced by Fossati et al. (1998) is its observational bias towards flaring sources in the VHE γ-ray regime. This bias has a strong influence on the correlation parameters found for compiling a spectral sequence, i.e. the correlation of the synchrotron and inverse Compton peak fluxes and frequencies. One can get rid of this bias by observing blazars in steady VHE emission states. Considering the relatively low number of blazars detected in the VHE regime so far, the number count for sources detected in steady-state emission is even lower. The observation of the steady state of blazars is also important to constrain the different emission models. In case of a single-zone SSC model a stationary shock within the jet is needed to maintain a steady emission. Marscher et al. (2008) discuss such a scenario for the case of BL Lacertae. Flaring states are then caused by perturbations moving along the jet and crossing the stationary shock.

1.2.3 BL Lac objects

BL Lac objects can be subdivided into low frequency peaked (LBL), intermediate (IBL) and high frequency peaked (HBL) BL Lac objects according to their synchrotron peak frequency lying at energies below $10^{14.5}$ Hz, at $sim 10^{14.5-16.5}$ Hz and above $10^{16.5}$ Hz, respectively (Nieppola et al., 2006). In the following the blazar subclass of HBLs will be brought into a short focus.

Characteristics

For observations above 100 GeV HBLs are particularly interesting due to their high synchrotron peak frequency leading to a measurable flux in the VHE regime. Due to that they represent the main VHE γ-ray emitting class. Among 34 extragalactic objects detected in the VHE regime 29 are blazars and 24 belong to the class of HBLs.

Variability

BL Lac objects exhibit variability on all wavelength- and timescales from years down to minutes. The reason could be shock waves moving along the relativistic jet and accelerating particles which lead to an enhancement of the VHE flux. The detection of LBLs was up to now only possible in states of high activity of the source whereas HBLs were also detected in phases with no or weak activity. The state of lacking any activity referred to as the baseline emission or steady state is of particular interest.

1.3 Blazar unification by evolution

Unifying blazars by means of their cosmological evolution was proposed by several authors. When characterising the evolution in general one has to distinguish between luminosity and density evolution which usually behave differently with time. Cavaliere and Malquori (1999) described the evolution of BL Lac objects. In contrast to the rest of AGN BL Lac objects exhibit only a weak evolution in luminosity as well as in density (cf. Stickel et al., 1991; Wolter et al., 1991; Bade et al., 1998). Evolutionary scenarios for blazars in general were proposed by D'Elia and Cavaliere (2001), Cavaliere and D'Elia (2002) and Böttcher and Dermer (2002). According to that the observational sequence FSRQ → LBL → HBL is not only an orientation effect (Fossati et al., 1998), but FSRQs and BL Lac objects represent different stages of development of the same source population showing a spectral hardening with decreasing luminosity and a reduction of the accretion power onto the central black hole with time. Böttcher and Dermer (2002) modeled this behaviour by changing the accretion rate or the optical depth of the circumnuclear material.

Responsible for the evolution and transition from FSRQs to BL Lac objects are the accretion of matter onto the central black hole and the density of the surrounding gas. In FSRQs with luminosities $L \gtrsim 10^{47}\,\mathrm{erg\,s^{-1}}$, strong accretion $\dot{m} \sim 1$ close to or above the Eddington limit takes place. The Eddington luminosity is defined as $L_{\mathrm{Edd}} = 1.26 \cdot 10^{47} M_9\,\mathrm{erg\,s^{-1}}$ with $M_9 = m/(10^9 M_\odot)$ being the mass of the black hole. $\dot{m}$ is then defined as (Boldt and Ghosh, 1999)

$$\dot{m} = c^2 \frac{\frac{\mathrm{d}m}{\mathrm{d}t}}{L_{\mathrm{Edd}}} \ . \tag{1.1}$$

This rate can only be maintained if enough gas is available for feeding the accretion disk and the black hole. Therefore the observed emission of FSRQs is dominated by the thermal Big Blue Bump in their SEDs. Additionally the rich environment leads to BLRs and enables the production of external Compton emission in leptonic scenarios. The stockpile of gas in the central region can be refilled by interactions of the host galaxy with companions in merging events. Accretion is then triggered by these events on timescales of $\tau \sim \mathrm{O}(10^8\,\mathrm{a})$. This procedure can repeat 3-5 times after $z \sim 2.5$. As there are more merging events at higher redshift and thus

1 Evolution of blazars

the frequency of such events decreases with time, the density evolution of FSRQs is weak and positive with timescales $\tau_D \sim (5-7) \cdot 10^9$ a. However, a strong luminosity evolution can be expected because the efficiency of each interaction and following accretion episode becomes less due to the exhaustion of host galaxy gas caused by previous accretion episodes or star formation. The timescale for the luminosity evolution is $\tau_L \sim 3 \cdot 10^9$ a (Cavaliere and D'Elia, 2002).

BL Lacs arise in this scenario from FSRQs after a last interaction event. The circumnuclear gas is thinned out and the accretion onto the black hole decreases significantly to rates $\dot{m} \sim 10^{-2} - 10^{-3}$ in Eddington units. The power source fueling the luminosity $L \lesssim 10^{46}\,\mathrm{ergs}^{-1}$ is now mainly non-thermal with partial contributions from the disk and the black hole. The latter one powers the jet with rotational energy extracted by means of the Blandford-Znajek mechanism. A detailed description is given in Blandford and Znajek (1977). This process can be maintained over long timescales because the black hole could gather enough angular momentum during the phases of accretion in the FSRQ state. Thus the luminosity evolution of BL Lacs is positive but weak with timescales $\tau_D \sim (5-10) \cdot 10^9$ a. The density evolution mirrors the corresponding evolution of FSRQs, except that it is negative, so $\tau_D \sim -(5-7) \cdot 10^9$ a.

Table 1.3 gives a summary of the important values concerning the evolution of blazars (cf. Cavaliere and D'Elia, 2002). Listed are also the comparison of the luminosities of the central black hole L_{BH} and the disk L_D as well as the top photon energies. However, the latter values have to be taken with caution, as the discovery of VHE emission from the FSRQ 3c279 clearly extended the value given in the table (Albert et al., 2008e). Thus this statement gives only a rough estimation of the photon energies found in these sources. Additionally one has to distinguish between active (flaring) and baseline states.

Continuing the evolutionary scenario to sources with even less accretion rates and a dominating disk luminosity, one can expect the acceleration of UHE cosmic rays within the jets of these quasar remnants. This topic will be discussed in chapter 2.

1.3 Blazar unification by evolution

Parameter	FSRQs	BL Lac objects
Key parameter	$\dot{m} \sim 1$	$\dot{m} \sim 10^{-2}$
Optical features	Emission lines, Big Blue Bump	No or weak lines, no bump
Integrated power	$L \sim 10^{47}\,\mathrm{erg\,s^{-1}}$	$L \lesssim 10^{46}\,\mathrm{erg\,s^{-1}}$
Black hole power vs disc power	$L_{BH} \ll L_D$	$L_{BH} \lesssim L_D$
Top energies	$\sim 10\,\mathrm{GeV}$	$\sim 10\,\mathrm{TeV}$
Evolution	Strong	Weak, if any
Timescales	$\tau_D \sim (5-7) \cdot 10^9\,\mathrm{a}$ $\tau_L \sim 3 \cdot 10^9\,\mathrm{a}$	$\tau_D \sim -(5-7) \cdot 10^9\,\mathrm{a}$ $\tau_L \sim (5-10) \cdot 10^9\,\mathrm{a}$

Table 1.3: Key parameters and values characterising the evolutionary states of blazars: FSRQs and BL Lac objects. Adapted from Cavaliere and D'Elia (2002).

1 Evolution of blazars

2 The very high energy γ-ray and cosmic ray connection

In this chapter an overview on cosmic rays is given followed by a description of the production and propagation of VHE γ-rays. The connection between both can be drawn by considering the sources capable of accelerating the particles to very or ultra high energies.

2.1 Cosmic rays

In 1912 Viktor Hess measured an increase of ionising radiation with increasing height in the atmosphere. This was in contrast to the general expectation that the main radioactive radiation was coming from the ground. The ionising radiation, later called 'cosmic radiation' by Robert Millikan, consists of massive charged particles such as protons, alpha particles, heavier nuclei and electrons as well as photons impinging permanently on the earth's atmosphere.

Figure 2.1 shows the cosmic ray spectrum between $\sim 1\,\text{GeV}$ and $10^{20}\,\text{eV}$. It is characterised by a steep power law shape, exhibiting some features to be investigated in more detail:

Knee
At the so-called knee around $4.6 \cdot 10^{15}\,\text{eV}$ the spectrum steepens from an index $\alpha \sim 2.7$ to $\alpha = 3$. The reason presumably is that at higher energies protons produced within the Milky Way are not confined anymore to our galaxy. This assumption is supported by the fact that at an energy of $4.6 \cdot 10^{15}\,\text{eV}$ the gyroradius of particles like protons exceeds $5\,\text{pc}$ – assumed to be the escape condition from the galactic disc – in a galactic magnetic field of $10^{-10}\,\text{T}$.

Ankle
At an energy of $\sim 5 \cdot 10^{18}\,\text{eV}$ the spectrum flattens again. There exist different models to explain the characteristics of the knee and the ankle concerning the transition from galactic to extragalactic cosmic rays (cf. for instance Wibig and Wolfendale, 2004; Aloisio et al., 2007). Above the ankle the cosmic rays are most probably of extragalactic origin.

2 The very high energy γ-ray and cosmic ray connection

GZK-cutoff

At $\sim 6 \cdot 10^{19}$ eV the cosmic rays interact with the photons of the 2.7 K cosmic microwave background (CMB). This leads to a cutoff in the spectrum. Only UHE cosmic rays within a distance of ~ 75 Mpc can reach the earth before absorption within the CMB.

The most important acceleration process for cosmic rays is Fermi acceleration. Charged particles gain energy in diffusive shock acceleration processes by crossing multiple times either a shock front or plasma waves. Due to the energy gain proportional to the velocity $\beta = v/c$ of the particle the former acceleration process is called first order Fermi acceleration. The latter one in analogy is second order Fermi acceleration, because of the energy gain of the particle proportional to β^2.

Possible candidates for cosmic ray acceleration were proposed by Hillas (1984). In order to escape from an accelerating region a cosmic ray has to exceed twice the larmor radius. This condition – depending on the atomic number of the accelerated particle – sets limits on the ratio of the size of and the magnetic field within a production site. Figure 2.2 shows the size vs. the magnetic field of an accelerating region. Additionally the different magnetic field conditions depending on the size for protons and iron nuclei with energies of 10^{20} and 10^{21} eV are shown.

The reason for not being able to identify directly sources of cosmic rays is that the charged particles are deflected in galactic and intergalactic magnetic fields. Only cosmic rays with the highest energies around 10^{20} eV are not influenced by magnetic fields due to their high rigidity $R = E/(Ze)$. But astronomy at these energy scales is restricted to close-by sources up to ~ 100 Mpc due to the GZK-cutoff.

2.1 Cosmic rays

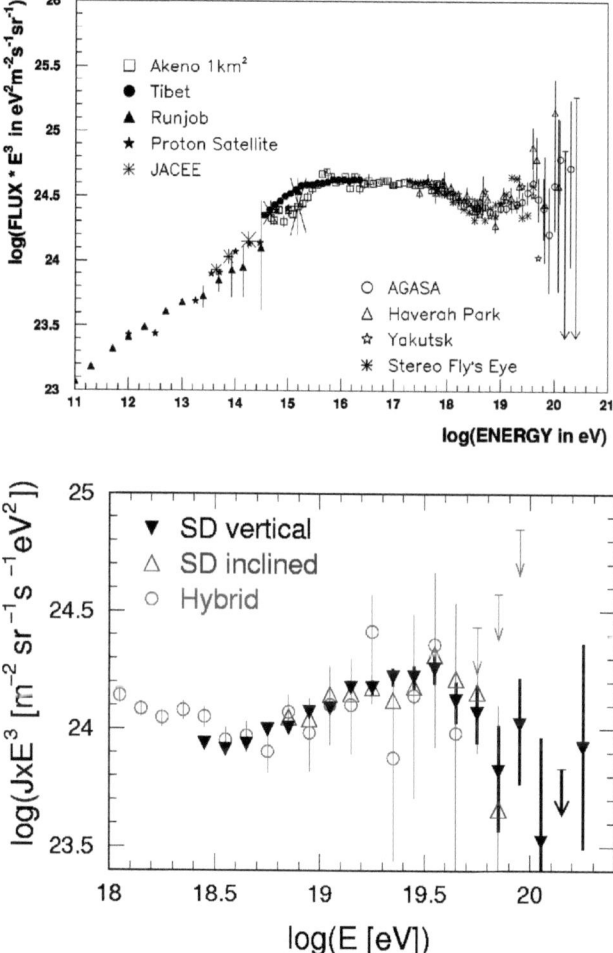

Figure 2.1: Top panel: Cosmic ray spectrum. It is described by a steep power law which is modified at the knee and ankle. Taken from Nagano and Watson (2000). Bottom panel: The UHE cosmic ray spectrum as measured by the Auger Collaboration. The ankle and the region around the GZK-cutoff are clearly visible. Cf. Yamamoto (2008) for further details.

2 The very high energy γ-ray and cosmic ray connection

2.2 Very high energy γ-rays

Contrary to cosmic rays, γ-rays being photons do not lose their direction information. Thus they can be used for the investigation of sources in the sky. Additionally the measured spectra can help in unveiling the acceleration mechanisms inside the sources. There are different terminologies for the definition of X- and γ-rays: In atomic and nuclear physics γ-rays are connected to interactions within the nuclei, whereas X-rays emerge from processes of the atomic shell. In astrophysics the energy is the defining characteristics for X- and γ-rays. γ-rays start at an

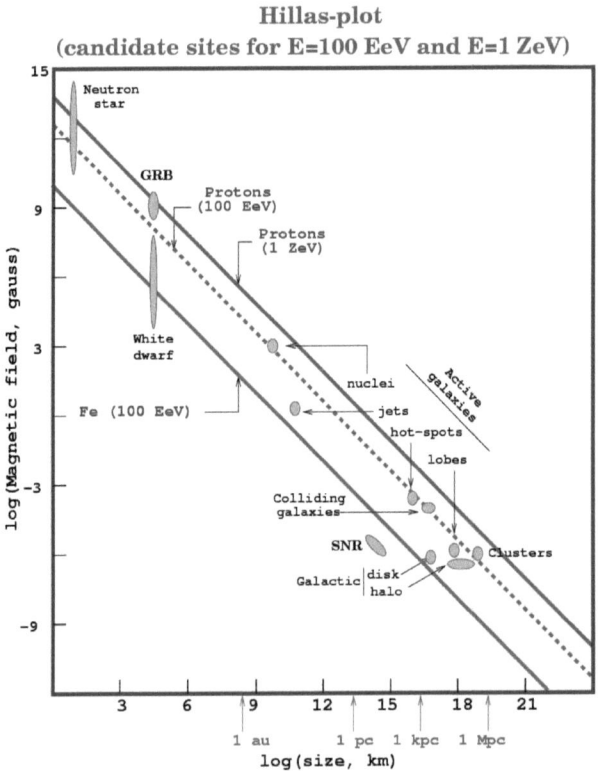

Figure 2.2: Hillas plot. Shown are different possible production regions for cosmic rays in a plane magnetic field vs size of the region. The production conditions for protons with energies of 10^{20} and 10^{21} eV as well as for iron nuclei with 10^{20} eV are plotted as red and green lines, respectively (www-02).

26

2.2 Very high energy γ-rays

energy of $\sim 1\,\mathrm{MeV}$. The subdivision of γ-rays into high energy (HE), very high energy (VHE) and ultra high energy (UHE) is not strictly defined. In this work the regime of VHE γ-rays ranges from $\sim 30\,\mathrm{GeV}$ to $100\,\mathrm{TeV}$.

2.2.1 Production mechanisms

The most energetic photons produced thermally in the universe reach up to X-ray energies of a few keV. VHE γ-rays are generated in non thermal production mechanisms which can be divided into electromagnetic (inverse Compton effect and Bremsstrahlung) and hadronic (π^0 decay) production.

π^0 decay

In nucleonic such as proton-nucleon or proton-photon interactions neutral and charged pions $\pi^{0,\pm}$ are produced. While charged pions have a long lifetime of $\sim 2.6 \cdot 10^8\,\mathrm{s}$, the neutral pions decay almost immediately into two γ-photons with a mean lifetime of $\sim 10^{-16}\,\mathrm{s}$.

Inverse Compton scattering (IC)

The interaction of a photon scattered off an electron at rest is called Compton scattering. In case of a relativistically moving electron the photon gains energy and the process is hence called inverse Compton scattering. The cross section for this process is given by the Klein-Nishina formula:

$$\sigma_{\mathrm{KN}} = \frac{3\sigma_{\mathrm{T}}}{4}\left\{\frac{1+x}{x^3}\left[\frac{2x(1+x)}{1+2x} - \ln(1+2x)\right] + \frac{\ln(1+2x)}{2x} - \frac{1+3x}{(1+2x)^2}\right\} \quad (2.1)$$

with $x = (E_\gamma \gamma_L)/(m_e c^2)$ and γ_L the Lorentz factor.
In case of $E_e E_\gamma \ll m_e^2 c^4$ the formula simplifies to the classical Thomson cross section

$$\sigma_{\mathrm{T}} = \frac{8}{3}\pi r_e^2 \,. \quad (2.2)$$

In case of ultrarelativistic electrons ($x > 10$) the approximation

$$\sigma_{\mathrm{KN}} = \pi r_e^2 \frac{\gamma_L}{x}\left[\ln\left(\frac{2x}{\gamma_L}\right) + \frac{1}{2}\right] \quad (2.3)$$

can be used. In the Klein-Nishina regime the electron looses large parts of its kinetic energy up to the total energy in one scattering process.

2 The very high energy γ-ray and cosmic ray connection

Bremsstrahlung

Bremsstrahlung is emitted, when a particle is deflected in an electric field. It is one of the most important processes in air shower physics (cf. section 3.1).

Synchrotron radiation

In the presence of magnetic fields ultrarelativistic electrons emit synchrotron radiation. The peak emission energy is given by
$E_{\text{peak}} = 5 \cdot 10^{-9} B_{\perp,G} \gamma_L^2$ eV with $B_{\perp,G}$ the magnetic field component in Gauss perpendicular to the electron's motion.

2.2.2 Intergalactic absorption of γ-rays

The free propagation of VHE γ-rays is limited by pair production with photons of the extragalactic background light (EBL):

$$\gamma_{\text{VHE}} + \gamma_{\text{EBL}} \rightarrow e^+ + e^- \quad (2.4)$$

with $E_{\gamma_{\text{VHE}}} \cdot E_{\gamma_{\text{EBL}}} > 2(m_e c^2)^2$ and a maximum of the pair creation probability around $2(m_e c^2)^2$. The minimum wavelength of the low energy photons is given by

$$\lambda_{\text{EBL}}(E_{\gamma_{\text{VHE}}}) = hc \frac{E_{\gamma_{\text{VHE}}}}{2(m_e c^2)^2} \sim 2.4 \frac{E_{\gamma_{\text{VHE}}}}{\text{TeV}} \mu\text{m} \ . \quad (2.5)$$

Therefore photons measured in the VHE γ-ray regime between $\sim 100\,\text{GeV}$ and $\sim 10\,\text{TeV}$ are interacting with infrared photons with wavelengths between $0.24\,\mu\text{m}$ and $24\,\mu\text{m}$.

2.3 Sources

Both populations, UHE cosmic rays and VHE γ-rays, are originating from very extreme energetic processes implying the capability inside the sources to provide these high energies. As the interest of this work is lying on the investigation of VHE γ-radiation from blazars, the main focus will be extragalactic sources. For radiation coming from outside the Milky Way many source candidates come into consideration which provide the possibility to account for the acceleration of both species.

Albeit extensive air showers were already detected in the 1960s, the VHE astronomy only started in 1989 when the first source was discovered by Whipple: the Crab Nebula. By that time, the field of HE γ-ray astronomy experienced a significant upturn in 1990 with the launch of

2.3 Sources

EGRET (Energetic Gamma-Ray Experiment Telescope) on board of CGRO (Compton Gamma Ray Observatory). Above an energy of 20 MeV EGRET discovered 271 sources (Hartman et al., 1999). A revision of the EGRET data using a different model for the diffuse γ-ray emission resulted in a new catalogue with only 188 sources (Casandjian and Grenier, 2008). At higher energies the effective area of satellite experiments becomes too small to achieve reasonable count rates because the γ-ray energy spectra are steeply decreasing. However, ground-based Cherenkov telescopes are best suited to cover the VHE domain above $\sim 10\,\mathrm{GeV}$. With the launch of the Fermi Gamma-ray Space Telescope (FGST) in 2008 the energetic gap between space-based and ground-based experiments is closed now. It serves as an effective trigger for promising sources to be observed in the VHE regime (cf. for instance Abdo et al., 2009).

Cosmic rays also produce extensive air showers in the atmosphere. In fact they represent the much more abundant background for imaging atmospheric Cherenkov γ-ray telescopes. As already explained above and can be seen in figure 2.1 extragalactic cosmic rays presumably dominate the spectrum above the 'ankle' at $\sim 10^{18-19}\,\mathrm{eV}$. Thus the following list concentrates on UHE cosmic ray and VHE γ-ray sources.

2.3.1 γ-ray bursts

The most extreme objects among astrophysical sources are γ-ray bursts (GRB). They produce an energy output of up to $10^{54}\,\mathrm{erg\,s^{-1}}$ within seconds to some tens of seconds. As they are distributed isotropically in the sky, they are assumed to have an extragalactic origin. The common model of a GRB is the fireball model, where a compact progenitor produces an ultrarelativistic outflow of plasma which is optically thick. The GRB is released in the very moment the plasma gets optically thin. As progenitors several scenarios are under discussion: Two merging compact objects like neutron stars or black holes or the death of very massive stars (hypernova or collapsar). Figure 2.3 shows an artist view of a possible scenario of two merging neutron stars. The duration of GRBs ranges from $\sim 10^{-3}\,\mathrm{s}$ to $\sim 10^{3}\,\mathrm{s}$ and is twofold with a separation at $\sim 2\,\mathrm{s}$. The observation of GRBs is one key goal of the MAGIC telescope. Therefore the telescope structure and drive system were designed to assure fast repositioning times of some tens of seconds. The MAGIC telescope will be presented in detail in section 3.4.

The generation of UHE cosmic rays in GRBs was proposed by Vietri (1995). The particles are either Fermi-accelerated at hyperrelativistic shocks or produced bouncing off the fireball in its freely expanding phase.

2 The very high energy γ-ray and cosmic ray connection

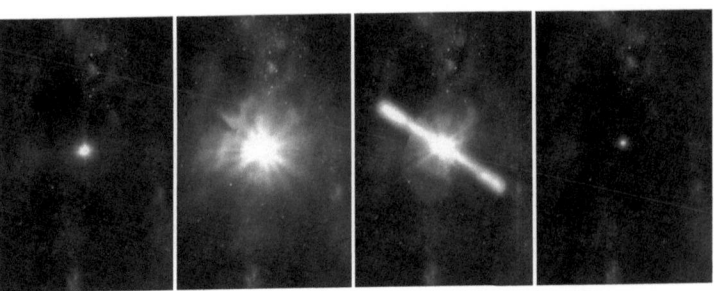

Figure 2.3: Artist view of a possible GRB scenario with two merging neutron stars. From left to right the merging event leads to a fireball and an ultrarelativistic plasma outflow in the form of two jets. Picture credit: NASA/D. Berry.

2.3.2 Starburst galaxies

In so-called starburst galaxies the star formation rate and thus also the number of massive stars and the supernova rate is high. The remainders of the star explosions – supernova remnants – are one of the most important sources for galactic cosmic rays. Therefore one expects a high abundancy of cosmic rays in these galaxies. The interaction of galactic cosmic rays with the stellar winds from the massive stars leads to the production of VHE γ-rays. Recently two starburst galaxies were discovered in the VHE γ-ray regime (Acero et al., 2009; Acciari et al., 2009b). In turn, the cosmic rays can be reaccelerated to energies well above $\sim 10^{21}\,\text{eV}$ at the terminal shock of the galactic superwind (Anchordoqui et al., 1999). Figure 2.4 shows a color-coded picture of the starburst galaxy M82 with the horizontal stellar disc and the galactic superwind of ionised gas perpendicular to it.

2.3.3 Compact objects

Arons (2003) suggested young rapidly rotating magnetars in nearby galaxies ($d \lesssim 50\,\text{Mpc}$) as a possible origin of UHE cosmic rays above the GZK cutoff. Magnetars are supernova remnants of massive stars in the form of pulsars with magnetic fields on their surface of order 10^{15} G. They occur in all galaxies where star formation takes place and could accelerate light ions up to energies of $\sim 10^{21-22}\,\text{eV}$. An artist view of a magnetar provides figure 2.5.

2.3 Sources

Figure 2.4: The starburst galaxy M82. The color-coded picture shows the horizontal stellar disc. The starburst inside the disc powers a perpendicular galactic superwind of ionized gas (purple). Picture credit: M. Westmoquette (University College London), J. Gallagher (University of Wisconsin-Madison), L. Smith (University College London), WIYN/NSF, NASA/ESA.

2.3.4 Active galactic nuclei

AGN belong to the most powerful objects in the universe with luminosities up to $10^{47}\,\mathrm{erg\,s^{-1}}$. As already discussed in section 1.1, AGN consist of a supermassive black hole with a mass of 10^6 to $10^{10}\,\mathrm{M_\odot}$, surrounded by an accretion disc and a dusty torus. Perpendicular to the accretion disc a relativistic matter outflow from the black hole region builds up a jet where a continuum photon emission from the radio up to the γ-ray domain is produced. Figure 2.6 shows an artist picture of an AGN.

Radio galaxies

With M87 and Centaurus A the two closest giant radio galaxies have been established as VHE γ-ray emitters (Aharonian et al., 2003a, 2009). Due to the relatively low angle of $10° - 19°$ between the jet axis of the AGN in M87 with the line of sight of the observer, this radio galaxy is often referred to as a 'misaligned' blazar. This statement is supported by the fact that in the past the detection of γ-rays was assigned to the so-called knot HST-1 inside the jet. On the other hand recent multiwavelength observations indicate that also the nucleus could be the

2 The very high energy γ-ray and cosmic ray connection

Figure 2.5: Artist view of a magnetar with the magnetic field and a rotating radio jet. Picture credit: John Rowe Animations.

origin of VHE γ-ray emission (Wagner et al., 2009a). Radio galaxies are also assumed to be emitters of UHE cosmic rays, supported by the detection of the latter ones with the Pierre Auger observatory. 27 cosmic ray events above an energy of $5.7 \cdot 10^{19}$ eV were detected and the origin of 20 among them was correlated to AGN (Abraham et al., 2007, 2008). Two UHE cosmic ray events originated within $3°$ from Centaurus A serving as a tentative confirmation of the theory.

Blazars

The most important subclass for VHE γ-ray observations are blazars and in particular BL Lac objects due to their strongly beamed emission towards the observer. Up to now 32 AGN have been confirmed as VHE γ-ray emitters with 29 blazar identifications and 24 among them belonging to the subclass of HBLs. Within the jets of AGN also protons and heavier nuclei could be accelerated to energies high enough to escape the AGN and become cosmic rays detectable on earth. Due to the GZK limit cosmic ray astronomy is only possible for sources being located inside a sphere with radius $\sim 50 - 100$ Mpc and producing cosmic rays with high enough energies to prevent deflection on intergalactic magnetic fields. Cosmic rays from sources at larger distances get 'decelerated' to energies below $\sim 10^{20}$ eV and isotropised due to magnetic field interactions.

2.3 Sources

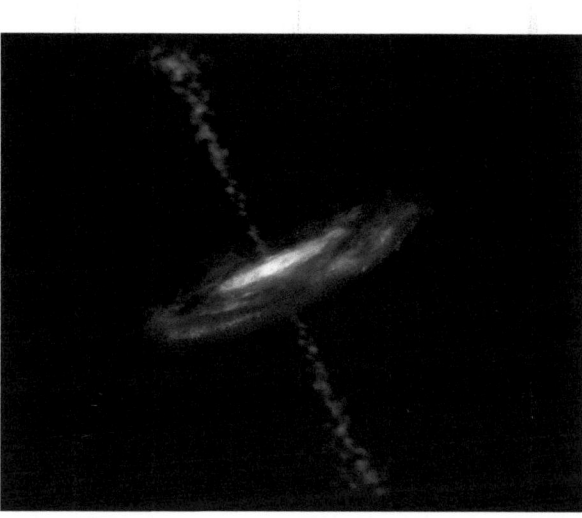

Figure 2.6: Artist view of an AGN showing the accretion disc, the dust torus and the relativistic jets. Picture credit: Aurore Simonnet, Sonoma State University.

Quasar remnants

In contrast to VHE γ-rays where the observer is still mostly restricted to the beamed emission of blazar jets, the cosmic ray isotropisation allows not only blazars but AGN in general to serve as possible sources. One possible scenario for cosmic ray production in an AGN is provided by Boldt and Ghosh (1999), Cavaliere and D'Elia (2002) and Isola et al. (2003). Continuing the evolution from FSRQs to BL Lac objects as described in chapter 1 leads to sources with luminosities $L \lesssim 10^{42}\,\mathrm{erg\,s^{-1}}$ even lower than the ones found in BL Lac objects and totally dominated by the black hole luminosity L_{BH}. The accretion rate has decreased to $\dot{m} \sim 10^{-4}$. These quasar remnants could accelerate massive particles by means of an electromotive force inside the jets to ultra high energies of $\sim 10^{21}$ eV.

2.3.5 Diffuse emission

As cosmic rays with energies less than $\sim 10^{18}$ eV or from distances larger than $\sim 100\,\mathrm{Mpc}$ are isotropised by intergalactic magnetic fields, the largest part of the extragalactic cosmic ray spectrum consists of diffuse emission. In the HE and VHE γ-ray regimes this is different because photons keep their direction information. Therefore only unresolved sources contribute to the diffuse γ-ray background. The EGRET instrument and recently the FGST measured the extragalactic γ-ray background up to energies of ∼ 40 GeV (Strong et al., 2004; Inoue et al., 2010). Although the origin of the diffuse emission is not yet fully understood, blazars are assumed to account for a sizable part of ∼ 45% to the diffuse γ-ray flux at ∼ 100 GeV and more at higher energies (Inoue and Totani, 2009).

3 Imaging atmospheric Cherenkov technique

The earth's atmosphere is impervious to electromagnetic waves except for the energy bands in the optical and radio regime. In order to get information for other energy ranges, the detector has to be placed in higher altitudes or even outside the atmosphere. Figure 3.1 shows the opacity of the atmosphere for electromagnetic waves. It is obvious that for γ-rays including the VHE regime the atmosphere is not permeable, thus only satellites can detect this radiation directly. Due to their relatively low collection area for VHE γ-capture one avails oneself of the imaging atmospheric Cherenkov technique (IACT) with ground-based detectors. The main advantage of this technique is the usage of the atmosphere as a calorimeter offering a huge effective collection area. The γ-rays are absorbed in the atmosphere by interacting with nuclei of the upper atmosphere's gas. In this process an avalanche of secondary particles is triggered, producing an extensive air shower (EAS). The secondary massive particles are producing Cherenkov light detectable on the ground.

In the following the physics behind extensive air showers and the Cherenkov effect will be presented. Afterwards the imaging technique for detecting EASs is explained, followed by a description of the currently largest single-dish imaging atmospheric Cherenkov telescope: MAGIC.

3.1 Extensive air showers

When a γ-ray or a charged cosmic ray strikes the upper atmosphere, an extensive air shower is initiated. In case of γ-rays an electron-positron pair is created within the Coulomb field of a nucleus in the atmosphere. The $e^{+/-}$ produce again γ-photons by suffering from Bremsstrahlung which again produces $e^{+/-}$-pairs and so forth. In this way the air shower evolves creating 2^N particles after N radiation lengths. When the $e^{+/-}$ energies drop below $\sim 83\,\text{MeV}$, ionisation losses become dominant and the air shower dies out.

3 Imaging atmospheric Cherenkov technique

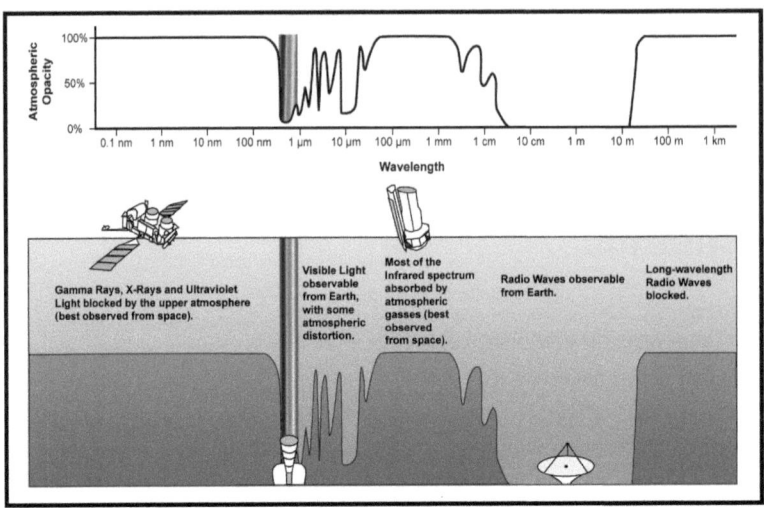

Figure 3.1: Opacity of the atmosphere for electromagnetic waves. Only in the optical and radio bands the atmosphere is transparent. In order to observe other wavebands one has to leave the atmosphere (e.g. with balloons or satellites) or make use of secondary particle detection techniques.

In case the impinging particle is a charged cosmic ray, the interaction with a nucleus leads to a larger variety of produced secondary particles. One can separate the shower qualitatively into three parts, a hadronic, a muonic and an electromagnetic one, depending on the particle species created in the different interactions. One of the key characteristics of both types of air showers is the different shape of the evolving particle cascades. This feature is the most important motivation for separating the two populations in imaging detectors as described in section 3.3. In figure 3.2 two simulated EAS are shown induced by a γ-ray and a cosmic ray proton, respectively, each with an energy of 100 GeV for the primary particle.

3.2 Cherenkov effect

Due to their relativistic energies the secondary particles in the EAS are travelling almost with the vacuum speed of light c_0, which is faster than the speed of light in the medium of the atmosphere, c_n.

A particle moving at the speed v through a medium is polarising the surrounding molecules. If

3.2 Cherenkov effect

the particle is faster than c_n, the polarisation is asymmetric. According to Huygen's principle the relaxation of the molecules then leads to a net electromagnetic wavefront moving away

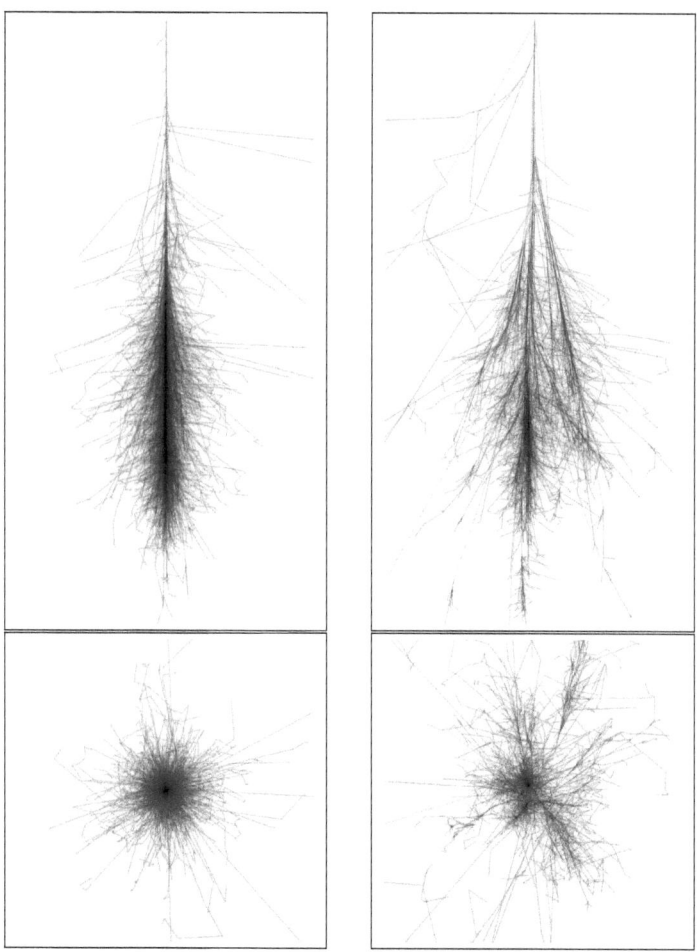

Figure 3.2: Left: CORSIKA-simulation of a γ-shower with 100 GeV energy of the primary particle. Right: CORSIKA-simulation of a proton shower with 100 GeV energy of the primary particle. The upper panels shows the vertical development, the lower panels the top view in the direction of the shower axis. Different colours represent different particle tracks: Red: Positrons, electrons and γ-rays; Blue: Hadrons; Green: Muons (www-03).

3 Imaging atmospheric Cherenkov technique

from the particle, because the elementary waves of each dipole interfere constructively. This Cherenkov effect is analoguous to a supersonic boom. The massive particles emit so-called Cherenkov radiation until v drops below c_n. Thus the condition for Cherenkov light emission is

$$v > c_n := \frac{c_0}{n}. \tag{3.1}$$

The opening angle θ_C of the Cherenkov light cone depends on the speed of the particle v and the refraction index n of the medium:

$$\cos\theta_C = \frac{1}{\beta n}, \quad \beta = \frac{v}{c_0} \tag{3.2}$$

The minimum angle – emission in forward direction – arises for $v_{\min} = c_n$ or $\beta_{\min} = 1/n$, which yields $\theta = 0°$. The minimum energy for particles to emit Cherenkov radiation follows as:

$$\beta_{\min} = \sqrt{1 - \left(\frac{E_0}{E_{\min}}\right)^2} = \frac{1}{n} \quad \text{with } E_0 = m_0 c^2 \tag{3.3}$$

$$\Rightarrow E_{\min} = \frac{1}{\sqrt{1 - \frac{1}{n^2}}} \cdot m_0 c^2 \tag{3.4}$$

Due to the dependence of $E_{\min}$ on the rest mass of the particle, lightweight particles as e^+ and e^- are the most important ones for the Cherenkov radiation. For calculating the maximum opening angle of the radiation cone and the minimal energy of the particle one has to know the refraction index which is dependent on the height in the atmosphere:

$$n(h) = 1 + \eta_0 \cdot e^{-\frac{h}{h_0}} \tag{3.5}$$

with $\eta_0 = 0.00029$ and $h_0 = 8400\,\text{m a.s.l.}$ The maximum refraction index in air is $n(\text{sea level}) = 1.00029$. For $\beta_{\min} = 1/n(\text{sea level}) = 0.99971$ follows $\theta_{\max} = 1.38°$. The mean height for EAS is $\sim 10\,\text{km}$, where $\theta_{\max}(10\,\text{km}) = 0.76°$. The diameter of the Cherenkov light cone on the ground is then $\sim 240\,\text{m}$. The energy that a particle is losing due to Cherenkov radiation along its path x is given by (Jackson, 1982):

$$\frac{dE}{dx} = \left(\frac{Ze}{c_0}\right)^2 \int_{(\beta n)^2 > 1} \left(1 - \frac{1}{(\beta n(\omega))^2}\right) \omega\, d\omega \tag{3.6}$$

with Z as the atomic number and ω the frequency of the Cherenkov photons. The integrand is the differential spectrum of the photons. It has its maximum at optical blue to UV frequencies.

3.3 Imaging technique

In contrast to a γ-ray satellite which uses a direct detection method for γ-rays, the imaging atmospheric Cherenkov technique used by ground-based telescopes is an indirect detection technique where the Cherenkov radiation produced by secondary particles in an EAS is recorded.

The collective Cherenkov radiation of all particles inside the EAS creates a light cone with $\sim 120\,\text{m}$ radius on the ground. An optical telescope inside this light pool can detect the EAS by reflecting the light into a camera. Due to the short time of the Cherenkov light flash of about 3 ns for γ-ray initiated cascades, the optical detectors are mostly photo multiplier tubes (PMTs) with a high timing resolution and fast readout electronics. One advantage of the IACT compared to satellite experiments is the huge collection area for air showers, which effectively is the area the light pool alights on the ground. For an EAS with 0° incidence angle (relative to the zenith), the effective area is $\sim 45200\,\text{m}^2$ and circular. For higher incident angles the area becomes larger and elliptical. On the other hand, the travelling path for the particles through the atmosphere is longer and thus only higher energy particles can be detected at higher zenith distances compared to perpendicular incidence. The primary particle energy is in first order proportional to the overall light content of the EAS.

3.4 MAGIC telescope

The MAGIC (Major Atmospheric Gamma-ray Imaging Cherenkov) telescope is situated atop the Roque de los Muchachos on the Canary island of La Palma at 2200 m a. s. l. (N 28°45', W 17°53'). Due to its geographical latitude it is especially suited for observations of extragalactic objects. For a zenith distance of 45° the declination range of potential sources is about -17° $< \delta < 73°$. The MAGIC telescope was designed to measure VHE γ-rays between $\sim 100\,\text{GeV}$ and 10 TeV. With a special trigger setup observations are possible down to 25 GeV as successfully proved with the Crab Nebula pulsar (Aliu et al., 2008).

The main physics goals for key observation programs with MAGIC are the detection of GRBs and the understanding of acceleration processes in galactic and extragalactic sources as for instance SNRs or AGN. In the following sections the key components of MAGIC are discussed.

3.4.1 Structure and Reflector

With its 17 m diameter tessellated mirror the MAGIC telescope is currently the world's largest imaging air Cherenkov telescope. The mirror has a focal length of 17 m ($f/d = 1$) and consists of 247 panels with a total surface of 234 m^2. The 956 single mirrors are spherical, 50 x 50 cm, diamond grinded and have an aluminium sandwich structure with a SiO$_2$ surface layer coating. Four of them are mounted on one panel. The panels can be heated to prevent icing and dew accumulation. The whole reflector is parabolic in order to keep the timing information of the recorded showers. It is mounted on a lightweight space frame structure of carbon fibre tubes. The total weight of the moving mass including the camera is 64 tons. The purpose of this lightweight design is to achieve fast repositioning times for catching prompt emission of GRBs (cf. section 2.3). The telescope can be pointed to any position in the sky within 40 s. It is an alt-az mount with two 11 kW motors for the azimuthal and another one for the elevation movement. The pointing accuracy is (1.1 ± 0.7) arcmin. A disadvantage of the lightweight structure is its lack of rigidity compared to a heavy steel frame. Because of that the telescope is bending depending on the elevation and weather influences like wind and temperature. Additionally,

Figure 3.3: The MAGIC telescope on the Roque de los Muchachos in the Canary island of La Palma. It is currently the largest Cherenkov telescope worldwide.

3.4 MAGIC telescope

in the course of time the bending changes. In order to keep the pointed source focused in the camera centre, the mirror panels can be moved by an active mirror control (AMC). The AMC is using look-up tables for each step of 5° in elevation to readjust the focusing of the mirror. In addition to that a CCD camera is monitoring the sky region of the current pointing in order to correct for mispointing introduced by the bending. This starguider system is comparing the observed star positions with catalogue positions and measures the mispointing. The latter can then be corrected for in the analysis software.

3.4.2 Camera

In the focal plane of the mirror at 17 m distance the camera is located. It consists of 576 PMTs ordered in a hexagonal shape. There are two kinds of pixels: The inner camera contains 396 1" PMTs with an angular field of view (FOV) of 0.1°, the outer camera is equipped with 180 1.5" PMTs with a FOV of 0.2° each. As the PMTs are round, the transition onto the hexagonal shape is done by light collectors called Winston cones. The overall FOV of the camera therewith is 3.5°. The PMTs have an enhanced quantum efficiency by $\sim 25\text{-}27\,\%$ in the wavelength range of 350-470 nm. This is achieved by coating their entrance windows with a special lacquer. The central pixel contains a detector for contemporaneous optical coverage during pulsar observations.

The dynamic range of the PMTs allows the telescope to run successfully under moon and twilight conditions. Thereby the possible observation time can be increased by up to $\sim 50\,\%$ (Albert et al., 2007e).

3.4.3 Data acquisition and trigger system

The analogue signals from the PMTs are transferred via 162 m long optical fibres to the receiver boards. The main advantage of separating the recording of signals and the electronic readout is the weight saving inside the camera which allows for faster repositioning for GRB observations. Inside the receivers the signal is split. One part is fed into the trigger system, another part is going to a flash ADC (FADC)[1] system.

Trigger system

The trigger system includes 325 inner camera pixels with an overall diameter of $\sim 2°$. In a first step the signal is going to a discriminator. Above a certain threshold a digital signal is sent. The discriminators reject as a first stage fluctuations of the night sky background (NSB)

---
[1] ADC: Analog Digital Converter

3 Imaging atmospheric Cherenkov technique

or bright stars in the FOV, because the thresholds are adjustable for each single trigger pixel. In a second step the topology of the triggered pixels is evaluated. Signals above the threshold have to have fast coincidences in N neighbouring pixels with a default value of $N = 4$. The trigger rate is typically around 200 to 300 Hz, depending on the atmospheric conditions, the zenith distance and the NSB.

FADC system

In January 2007 the FADC system of the MAGIC readout chain was updated and the new system will be referred to as 'MUX FADC' system hereafter. As the data presented in this work were taken both before and after that date, both systems will be discussed.

300 MHz FADC system

In case a trigger is issued 30 time slices are written to a first in first out (FIFO) buffer. The charge, timing and trigger information for the triggered event is then available. The signals are artificially stretched from 2.2 ns to $\sim$ 6 ns in order to measure the structure of the light pulses. Then the signal is split, one part is amplified by a factor of ten (high gain), the other part is left as it is (low gain). In case the high gain saturates the low gain is also digitized. The signal then consists either of 30 high gain slices or 15 high gain and 15 low gain slices. The length of a time slice is 3.3 ns and a light pulse usually has a width of four to six slices. From signals without low gain the late high gain part is used for getting pedestals (cf. also section 3.4.4).

2 GHz (MUX) FADC system

The new FADC system has several advantages compared to the 300 MHz system. The signals do not have to be stretched anymore, the integration time is minimised and the noise caused by the NSB is reduced. Furthermore the timing structure of the showers can be resolved much better. The MUX FADC system digitizes 16 read-out channels consecutively by delaying the analog signals by 40 ns per channel with optical fibres. A trigger signal is produced by splitting the original signal (Goebel et al., 2008). These optical splitters have been installed in an earlier stage introducing a change in the 300 MHz FADC system. Because of that the analysis software had to be adapted which led for instance to a splitting of the data sample of the Crab Nebula used for a comparative analysis (cf. section 4.6).

3.4.4 Observation modes and file types

In the data acquisition three run types are foreseen:

3.4 MAGIC telescope

Pedestal (P) runs
A P run consists of 1000 randomly triggered events. They reflect the signal baseline which consists of the fluctuations of the NSB and the electronic noise from the readout chain. With a P run an initial set of pedestals is taken.

Calibration (C) runs
A C run consists of about 4000 events triggered by the calibration system which illuminates uniformly the camera with pulsed light of a certain frequency, usually in the UV around the maximum of the Cherenkov spectral distribution. With a C run an initial set of calibration constants is calculated (cf. section 4.1).

Data (D) runs
A D run consists of events triggering the telescope while tracking a source. For the 300 MHz FADC system the pedestals during D runs were updated by taking the late high gain signals. The calibration constants were continuously updated by taking 50 Hz additional interlaced calibration events. For the MUX FADC system the interlaced events are now split into 25 Hz calibration and 25 Hz pedestal events due to the discontinuation of the high/low gain structure. D runs are stored when a certain file size has been reached according to a certain number of events.

At the moment the telescope is taking data in two different observation modes:

On-Off mode
In the On-Off mode the telescope is tracking a source in the center of the camera. The Off data are taken best in a similar sky region, where no γ-ray source is expected, accounting for the NSB with the same weather conditions and zenith distance range as the On region to get an estimate of the background. The background region (Off) is then subtracted in the analysis from the On region to obtain the signal events from the source.

Wobble mode
In order not to be dependent on Off observations which cost time and may not reflect the same conditions as in On observations, the wobble mode is combining On and Off recordings within one observation. Hereby the source position is displaced by typically 0.4° from the center of the camera. The region on the opposite side of the camera center is called the anti-source position. From this region the background is estimated. To get a higher background statistics, the two regions at $+90°$ and $-90°$ are taken as additional Off regions. With wobble observations the On-Off match better concerning the underlying background. As the camera acceptance may be inhomogeneous, the source and anti-source

3 Imaging atmospheric Cherenkov technique

positions are swapped regularly during observation to compensate systematic effects like this. Therefrom this method is called wobble mode. Furthermore the wobble mode offers the best time coverage at the expense of a lower efficiency compared to the central camera position.

3.4.5 Monte Carlo simulations

Monte Carlo (MC) simulations are indispensable for the data analysis of VHE γ-ray observations. On earth there is no VHE γ-ray source available for calibrating the telescope. MC simulations are therefore used for the reconstruction of the primary γ-particle energy and for the separation of hadronic induced air showers. For this purpose γ-initiated air showers and their development inside the atmosphere are simulated with CORSIKA (Heck et al., 1998). An example for two simulated air showers is shown in figure 3.2. Furthermore the telescope response and the PMT camera are simulated in two additional software steps, called Reflector and Camera.

4 Analysis chain

The data recorded with the MAGIC telescope are analysed and processed with a modular software package called MARS CheObs Edition[1] (Modular Analysis and Reconstruction Software Cherenkov Observatory Edition, Bretz and Dorner, 2008; Moralejo et al., 2009)) based on the ROOT[2] framework. In Würzburg a data centre provides access to all data taken with the MAGIC telescope from 2004 to 2009.

In this chapter the analysis steps are presented. The first analysis steps are carried out using a fully automised processing routine (Dorner et al., 2005). This accounts for the signal extraction and calibration as well as the event image reconstruction and cleaning (sections 4.1 and 4.2). In section 4.3 the background determination and suppression is described. The calculation of an energy spectrum can be found in section 4.4. After a short description of lightcurves in section 4.5, an exemplary analysis of observations of the Crab Nebula is eventually presented in section 4.6.

4.1 Signal extraction and calibration

Within the standard software framework, the signal extraction and calibration are done by a program called CALLISTO (CALibrate LIght Signals and Time Offsets). The telescope subsystems like for instance the drive system or the weather station provide additional information, which are needed or useful in the following analysis steps and thus written to the resulting output file by MERPP (MERging and Preprocessing Program). Finally, unsuitable pixels have to be removed from the analysis.

---

[1]The version used for the analysis presented in this work is the release version Mars_V2-3, http://magic.astro.uni-wuerzburg.de/mars/
[2]http://root.cern.ch/drupal/

4 Analysis chain

4.1.1 Signal extraction

As denoted in section 3.4, the DAQ stores the signals in 30 FADC slices. For the MUX FADC system the number of slices originally was 80, but was reduced to 50 by cutting the first and last 15 slices. The pulse is reconstructed from these FADC slices by means of a dedicated signal extraction algorithm. Furthermore the signal usually contains a background which has to be subtracted. This pedestal signal is caused by the NSB, local sources of light, e.g. stray light, and the electronic noise of the readout chain itself. It is extracted from dedicated events (pedestal runs) and continuously from the late high gain part of Cherenkov signals where no low gain was read out. As there is no high and low gain for the MUX FADC signals, the pedestal level is updated by recording dedicated interlaced pedestal events during data taking with a frequency of 25 Hz.

In order to reduce the noise, only the charge of a Cherenkov pulse itself should be extracted from the signal. For this purpose different extraction algorithms can be used (Albert et al., 2008a):

- **Spline extractor.** The pedestal-subtracted charge in the FADC slices is interpolated by a cubic spline algorithm. The reconstructed signal then is either the spline maximum or the integral of the spline centered on the spline maximum. As the arrival time values of the signal either the position of the spline maximum or the pulse half maximum of the rising edge can be used.

- **Digital filter algorithm.** This algorithm uses an analytic pulse form to extract the charges and the timing from the pedestal-subtracted FADC slices.

The analysis chain in the Würzburg data centre is using the digital filter whenever possible. In case that the Cherenkov pulses cannot be extracted due to e.g. too early arrival times, the spline algorithm is used.

4.1.2 Calibration

The charge extracted from the FADC slices has to be converted into the number of photoelectrons produced inside the PMTs. For this purpose dedicated calibration events (calibration runs) are recorded. Additionally, the conversion factors, also called calibration constants, are updated by using interlaced calibration events during data taking with a rate of 50 Hz. For the MUX FADC system this number is reduced to 25 Hz in order to allow for continuous pedestal coverage (25 Hz).

4.1 Signal extraction and calibration

The following conversion methods are used. For more information see Doro (2004) and Gaug (2006).

F-factor method

This is the standard calibration method. The excess noise factors of the PMTs, also called F-factors, were measured for a sample of 20 PMTs before their installation. The mean number of photoelectrons is given by

$$\langle n \rangle_{\text{ph.el.}} = F^2 \frac{(\langle Q \rangle - \langle P \rangle)^2}{\sigma_Q^2 - \sigma_P^2}. \tag{4.1}$$

$\langle Q \rangle$ is the mean charge with its standard deviation σ_Q. $\langle P \rangle$ is the pedestal and σ_P its error arising from the NSB, electronic noise and extractor uncertainties. The excess noise factor is defined as

$$F = \sqrt{1 + \frac{\sigma_G^2}{\langle G \rangle}^2} \tag{4.2}$$

with the amplification of electrons in the PMT dynode system, $\langle G \rangle$ and its standard deviation σ_G. The F-factor has been measured for the PMT sample in the laboratory to $F = 1.15 \pm 0.02$. The caveat of this method is that the F-factors were only measured for a small number of PMTs and that the calibration constants are only relative, i.e. only the conversion from FADC counts to photoelectrons and not from photons to photoelectrons is calculated.

Muon ring analysis

Isolated muons produced close to the telescope – in height as well as in distance on the ground – are producing ring-like images inside the camera plane. Their energy can be compared to Monte Carlo simulated muons. The muon images are sensitive to different atmospheric conditions as well as a changing reflector performance. Thus an absolute light collection efficiency calibration is possible (Rose, 1995; Goebel et al., 2005).

The calibration constants obtained with the F-factor method are updated with results from the muon ring analysis on basis of observation periods[3].

4.1.3 Bad pixel treatment

Pixels are flagged as unsuitable or bad pixels, if the pixel value fluctuation is larger than 5 standard deviations compared to the average fluctuation of all pixels. This is the case for PMTs

---
[3]An observation period lasts from one full moon to the next being one month.

4 Analysis chain

illuminated by bright stars, pixels which could not be calibrated or for the missing central pixel. The missing signal of the bad pixel is then interpolated by means of an algorithm correlating the contents of one pixel with its neighbours (Rügamer, 2006). At least three surrounding pixels are required for interpolation.

4.2 Event image reconstruction

The next step in the analysis chain is the image cleaning and parametrisation. The program inheriting these topics is called STAR (STandard Analysis and Reconstruction).

4.2.1 Software trigger

The NSB fluctuations can generate random coincidences in the trigger logic. The main reason are small variations in the gain of the PMTs or arrival time fluctuations in the trigger logic. These artificial triggers are removed by simulating a trigger in the software acting on the calibrated signal which is corrected for these fluctuations. A minimum of 4 neighbouring pixels with a threshold of at least 5 photoelectrons within a time window of ± 0.5 time slices is required to form a software triggered event. In contrast to the hardware trigger the software trigger is applied to the events after the subtraction of the pedestals. The advantage is the additional suppression of events arising from background fluctuations. The further analysis time is decreased and the resulting trigger rate, now unbiased by random coincidences, becomes a good quality measurement of the atmospheric conditions.

4.2.2 Image cleaning

When recording an image of a Cherenkov shower, part of the pixels will be dominated by the Cherenkov signal, the rest by the NSB fluctuations. In order to get rid of the latter pixels, an algorithm removing the noise from the pixels is applied:

1. **Core/Boundary pixel determination**
   In a first step a pixel is assigned a so-called core pixel if its charge content exceeds a pre-defined threshold value of 6 photoelectrons. In addition pixels above a second threshold of 3 photoelectrons are called boundary pixels. Solitary core pixels and pixels not neighbouring a core pixel are removed.

2. **Arrival time coincidence of three pixels**
   The second step considers the time differences between the surviving pixels. The pixels

4.2 Event image reconstruction

containing a signal from the Cherenkov shower should be correlated in the arrival time, whereas pixels dominated by NSB fluctuations have uncorrelated arrival times. Each pixel not having at least two neighbouring pixels within a time coincidence window of 1.75 ns is removed.

3. **Arrival time coincidence of two pixels**
   The second step is repeated, but only requiring one neighbouring pixel within a time coincidence window of 1.75 ns.

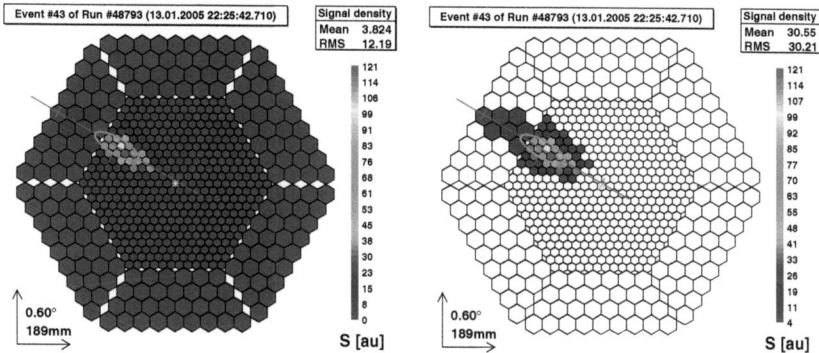

Figure 4.1: Calibrated (left panel) and cleaned (right panel) image of a γ-like EAS in the MAGIC camera plane.

4.2.3 Image parametrisation

In order to analyse the statistical properties of the cleaned shower images they are parametrised. The main parameters are the light content and the moments along the main axes of the shower image. Following Hillas (1985), the classical image parameters are:

width w: The second moment of the intensity distribution along the minor axis of the shower.

length l: The second moment of the intensity distribution along the major axis of the shower.

conc: The concentration. It is the ratio of the charge content in the two brightest pixels and the *size* (see below)[4].

---
[4] Even if not used directly in the analysis, similar parameters have been developed (e.g. concCOG and conc1) based on the concentration parameter.

4 Analysis chain

dist d: The distance between the center of gravity of the shower image and the source position in the camera plane.

α: The angle between the main shower axis and the axis connecting the source position with the center of gravity of the shower in the camera plane[5].

Additionally, the following parameters are used in the further analysis:

size s: The total charge content of the shower image.

area A: The area of the shower. It is defined as $\pi \cdot w \cdot l$.

m3long: The third moment of the intensity distribution along the major axis of the shower.

leakage1 $L1$: The ratio of the charge content of the shower in the outermost ring of camera pixels and the *size*.

slope: The time development of the shower along the major axis.

disp p: The distance between the center of gravity of the shower image and the shower origin. Introduced by Lessard et al. (2001), the disp method is used to reconstruct the shower origin. It is defined as

$$p = \xi \cdot \left(1 - \frac{w}{l}\right) . \tag{4.3}$$

In order to better reconstruct showers truncated at the camera edge, the constant factor ξ is modified with the *leakage1*. Furthermore the *slope* and the *size* are taken into account:

$$\xi = c_0 + c_1 \cdot slope + c_2 \cdot L1 + k \cdot (c_3 \cdot (\log_{10} s - c_4)) \tag{4.4}$$

with $k = 0$ for $\log_{10} s \leq c_4$ and $k = 1$ otherwise. p denotes only a distance, but does not contain any information about on which side of the shower the origin is located. Usually the shower image in the camera has a head and a tail along the major axis. To determine the head and the tail, the third moment and the time development along the major axis are used. *m3long* is assumed to be positive in the direction of the shower. Allowing for small negative fluctuations, especially in low energy showers, the condition for *m3long* is defined as

$$m3long < c_5 . \tag{4.5}$$

---
[5]In Hillas (1985) instead of the angle α the perpendicular distance between the source position and the main shower axis is defined as MISS. α can easily be derived from that parameter.

The time development *slope* gives an additional condition:

$$slope < (d - c_6) \cdot c_7 \qquad (4.6)$$

ϑ: The absolute distance between the shower origin calculated with the *disp*-method and the source position in the camera. It is calculated as

$$\vartheta^2 = d^2 + p^2 - 2 \cdot d \cdot p \cdot cos(\alpha) \qquad (4.7)$$

As denoted in section 4.1.2, muons are used for an absolute calibration of the shower events. Additionally, the optical point-spread-function (PSF) can be measured which is a probe for the quality of the mirror alignment. Therefore the ring images of muons hitting directly the reflector are parametrised:

radius: The radius of the muon image circle.

arcwidth: The width of the muon image circle.

muonsize: The total light content of the muon image circle.

arcphi: The length of the muon image circle.

In the context of background suppression, quality cuts as described in section 4.3.1 are performed based among others on the following parameters:

numislands N_i: The number of isolated clusters contained in the shower image.

numusedpix N_u: The number of pixels surviving the image cleaning.

concCOG: The ratio of the charge content of the three pixels next to the center of gravity (COG) and the *size*.

conc1: The ratio of the charge of the brightest pixel and the *size*.

4.3 Background rejection

The next analysis step is done by GANYMED (Gammas Are Now Your Most Exciting Discovery). It includes the quality cuts as well as image parameter cuts for background suppression and determines the significance of a potential detection. The rejection of the background is based solely on the statistical properties of the shower images.

4 Analysis chain

4.3.1 Quality cuts

The quality cuts are supposed to remove non physical shower images, e.g. sparks occurring in single PMTs, or events classified unambiguously as background by conditions derived through comparison of the parameter distributions with Monte Carlo simulated γ-showers. Additionally showers not containing sufficient information are also removed.

$$N_i < 3 \tag{4.8}$$
$$N_u > 5 \tag{4.9}$$
$$L1 < 0.3 \tag{4.10}$$
$$l > -3.6 \cdot (\log_{10} s - 6.0)^2 + 70 \tag{4.11}$$
$$\log_{10} concCOG < -0.45 + 0.08 \cdot (\log_{10} s - 3.9)^2 \tag{4.12}$$
$$\log_{10} conc1 < -0.75 + 0.10 \cdot (\log_{10} s - 3.8)^2 \tag{4.13}$$

Equations 4.12 and 4.13 are only valid for $\log_{10} s < 3.9$ and $\log_{10} s < 3.8$, respectively.

4.3.2 γ – hadron separation cuts

The main cut separating γ-like from background showers is made in *area*:

$$A < c_8 \cdot (1 - c_9 \cdot (\log_{10} s - c_{10})^2) \tag{4.14}$$

The *area*-cut is optimised on data of the Crab Nebula for obtaining both a good significance of the signal as well as many excess events (cf. figure 5.2).

As γ-showers coming from a source have their shower origin close to the source position in the camera, the distribution of ϑ should show an increase at small values. Therefore a cut in ϑ is done. It is the most important cut, because it is separating the signal region from the background regions.

$$\vartheta < c_{11} \tag{4.15}$$

The excess in the signal region can be quantified by calculating the significance of the signal by using formula 17 from Li and Ma (1983). It is also taking into account the background events inside the signal region and thus is depending on the On-Off ratio which is fixed to 1/3 in case of wobble mode observations (one On and three Off regions).

4.4 Spectrum

From the excess events a spectrum can be calculated by assigning an energy to the events. This is done by means of a random forest algorithm. Thereafter the spectrum of the excess events is calculated with SPONDE (SPectrum ON DEmand). The differential energy spectrum is defined by the differential γ-ray flux

$$\frac{\mathrm{d}N}{\mathrm{d}E}(E) = \frac{\mathrm{d}N_\gamma}{\mathrm{d}E \cdot \mathrm{d}A_{\mathrm{eff}} \cdot \mathrm{d}t_{\mathrm{eff}}} \qquad (4.16)$$

with N_γ being the excess γ-events and $\mathrm{d}A_{\mathrm{eff}}$ and $\mathrm{d}t_{\mathrm{eff}}$ the effective collection area (cf. section 4.4.2) and observation time (cf. section 4.4.3), respectively.

4.4.1 Energy estimation

Only statistical image parameters are assigned to the shower events, but their energy cannot be measured directly. Monte Carlo simulations of air showers yield the same image parameters with the advantage of a known energy. Thus these simulations can be used to reconstruct the energy of the primary γ-particle hitting the atmosphere by using a random forest algorithm on the Monte Carlos. The most important parameter for estimating the energy is the *size*. Additional parameters used are the *dist*, the zenith distance zd, the *leakage1* and the *slope*. The resolution being the error of the energy estimation has a mean value of usually 20-25%. For higher energies above $\sim$ TeV it becomes better, for the lowest energies it becomes worse because the background rejection works better at high energies than at low ones. In the medium energy range between 200 GeV and 1 TeV it is stable at the values given above.

When estimating the energy of showers, some showers are assigned a wrong energy based on the image parameters. As only statistical properties of the showers are considered, the distributions are always binned. The resulting spectrum is also binned in energy, which leads to a spill over of events with wrongly assigned energy into neighbouring bins. Additionally the image cleaning, artificial size cuts or wrong MC simulations can cause spill over effects. The solution for this problem is the spill over correction. For each energy bin the estimated energy E_{est} of the Monte Carlos is compared to their true energy E_{true}. The ratio of both gives a spill over factor a_i by which the resulting energy content in each bin i has to multiplied.

$$a_i = \frac{E_{\mathrm{true}}}{E_{\mathrm{est}}} \qquad (4.17)$$

4 Analysis chain

4.4.2 Effective collection area

For the spectrum determination the effective collection area for showers, depending on the energy, has to be calculated. The area on the ground spanned by a circle or ellipse around the telescope axis with a radius of the maximum impact parameter[6] considered can be projected to the sky as maximum area from which showers can be detected. Most of the showers which are detectable reach the telescope within an impact parameter of up to 300 m for higher zenith angle observations. Thus the area can reach values up to $A_0 = 2.8 \cdot 10^5$ m. The effective collection area is then defined as the maximum area A_0, modified with the cut and trigger efficiency, i.e. the ratio of the numbers N_0 of originally simulated and N_C of remaining simulated Monte Carlo showers after cuts:

$$A_{\text{eff}} = A_0 \cdot \frac{N_C}{N_0} \tag{4.18}$$

4.4.3 Effective observation time

The effective observation time t_{eff} is defined as the time in which a given number of events n during an observation time t_0 would be recorded by an ideal detector. t_{eff} is given as the number of events divided by the event rate λ. The latter is determined by an exponential fit to the distribution of the time differences Δt of consecutive events during data taking:

$$t_{\text{eff}} = \frac{n}{\lambda} \tag{4.19}$$

$$\frac{dn}{dt} = n_0 \cdot \lambda \cdot e^{-\lambda t} \tag{4.20}$$

4.4.4 Energy spectrum

Based on formula 4.16 the spectrum can be calculated for each energy bin i. The differential flux per bin consist of the excess events $N_{\text{exc},i}$, multiplied with the spill over correction a_i, divided by the effective observation time t_{eff} and collection area $A_{\text{eff},i}$ and the energy ΔE_i of the bin i, being just the width of the very bin:

$$\frac{dN_i}{dE} = \frac{N_{\text{exc},i} \cdot a_i}{t_{\text{eff}} \cdot A_{\text{eff},i} \cdot \Delta E_i} \tag{4.21}$$

---

[6]The impact parameter is the distance between the centre of the mirror and the point on the ground, where the main shower axis is pointing to.

4.5 Lightcurves

Lightcurves reflect the activity and variability if any of a source during the observations. They are produced by calculating the integral flux above a certain energy, mostly the energy threshold of the analysis, for a certain timescale. Usually the data taken during a single night are combined to build one lightcurve bin to check for variability on diurnal timescales. Other binnings are possible, depending mainly on the overall flux measured from the source. In case of high fluxes, smaller bins are feasible to resolve shorter timescales as was applied in case of a γ-ray flare of Mkn 501 in 2005 (Albert et al., 2007d). For lower fluxes larger bins are reasonable to provide a significant signal in each bin. If the signal from a source is only weak lacking a significant signal in the whole dataset one can plot the excess and background event rates to get an impression of the variability of the source. The latter method was applied to the objects presented in chapter 5.

4.6 Observations of the Crab Nebula

In this section the results of the analysis of observations of the Crab Nebula are presented. The analysis was done due to different reasons: (i) Check the stability and reliability of the hard- and software framework throughout 4 cycles of blazar observations (cf. chapter 5); (ii) check the performance of the applied analysis chain and compare the results to the published ones in Albert et al. (2008f); (iii) the integral flux measurement from the Crab Nebula is needed for comparison with the results of the blazar analysis. In order to obtain comparable results for the Crab Nebula and the analysis of the BL Lacertae sample, exactly the same analysis steps have been applied to all data.

4.6.1 Data selection and automatic analysis

Due to changes in the hardware three data samples of the Crab Nebula have been selected. As already mentioned in section 3.4, the data acquisition system was upgraded two times. In June 2006 optical splitters have been introduced in the 300 MHz FADC system in order to prepare the upgrade to a 2 GHz FADC system in January 2007. In principle the data originating from periods with different hardware setups can be analysed together. However, the determination of the energy spectrum requires different MC datasets and therefore the division into three subsamples is justified. The data have been chosen to match the observational parameters of the BL Lac object sample presented in chapter 5 as much as possible.

4 Analysis chain

As already described in section 3.4.4, the data are stored as raw runs after a certain number of events has been reached. These runs are automatically combined to sequences based on e.g. the observation mode and the trigger settings etc. The sequences are then calibrated, merged with subsystem information and the image cleaning as well as the calculation of the image parameters are done.

After the last automatic step a quality selection is performed based on the event rate after the image cleaning, the weather conditions (humidity, cloudiness, number of stars correlated by the starguider) and the inhomogeneity (cf. section 5.3). The final datasets are listed in table 4.1 including already the results of the analysis described in the following.

Season	FADC system	t_eff [h]	zd	N_sig	N_bck	N_exc	σ	E_thr [GeV]
10/2005 – 03/2006	300 MHz w/o splitters	3.80	6 – 37	1178	209	967	36.0	165
09/2006 – 01/2007	300 MHz w/ splitters	8.07	7 – 43	2609	523	2086	51.0	165
02/2007 – 01/2008	2 GHz	7.25	8 – 30	2588	455	2133	53.5	165
10/2005 – 01/2008	Combined	19.12	6 – 43	6376	1188	5188	82.2	165

Table 4.1: Observations of the Crab Nebula used for comparison to the flux upper limits of the BL Lac object sample and proof of concept of the stacking method described in section 5.3.

4.6.2 Background rejection

The background rejection or γ – hadron separation is done by means of cuts in different image parameter distributions. The set of cuts used here is shown in table 5.5. It is the standard set used in the data centre and was optimised on Crab Nebula data from different observations than the ones presented here. The only cut parameter which is different in both analyses (BL Lac object and Crab Nebula samples) is the ϑ-cut which has a value of $\vartheta < 0.21$ in case of the Crab Nebula analysis. This is the standard value which has been changed for the BL Lac object sample due to the expected smaller and thus narrower ϑ^2-distributions allowing for a better background reduction. Figure 4.2 shows the ϑ^2-distribution for the combined Crab Nebula sample, the individual ϑ^2-plots can be found in appendix B.

4.6.3 Energy spectrum

In order to check the stability of the telescope regarding the hardware changes the energy spectra have been calculated for the three subsamples. The energy estimation was done using a

4.6 Observations of the Crab Nebula

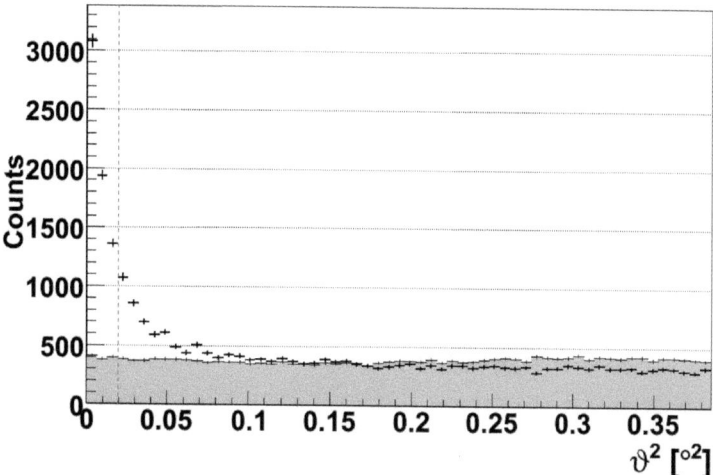

Figure 4.2: ϑ^2-distribution of the combined Crab Nebula data sample.

MC sample with a simulated energy spectrum corresponding to the published one for the Crab Nebula (Albert et al., 2008f). By changing two cut values (cf. table 5.5) the energy threshold of the analysis could be lowered to 165 GeV for each subsample.

In figure 4.3 the spectra of the subsamples as well as the combined spectrum are shown. They are not fully compatible within statistical errors, but systematic errors have not been taken into account. Integral fluxes will be compared to the combined spectrum later on. The spectra can be fitted with a parabolic function in a log–log representation following

$$\frac{dN}{dE} = f_0 \cdot \left(\frac{E}{E_0}\right)^{a+b\cdot\log_{10}(E/E_0)} \tag{4.22}$$

with f_0 being the differential flux in $(\text{TeV}\,\text{cm}^2\,\text{s})^{-1}$ and a the spectral slope at E_0 which was chosen here to $E_0 = 300\,\text{GeV}$. b is a measure for the curvature. Table 4.2 lists the results together with the parameters of the fit to the combined spectrum. As proof of concept the combined Crab Nebula analysis was performed with the stacking method described in section 5.3.

4 Analysis chain

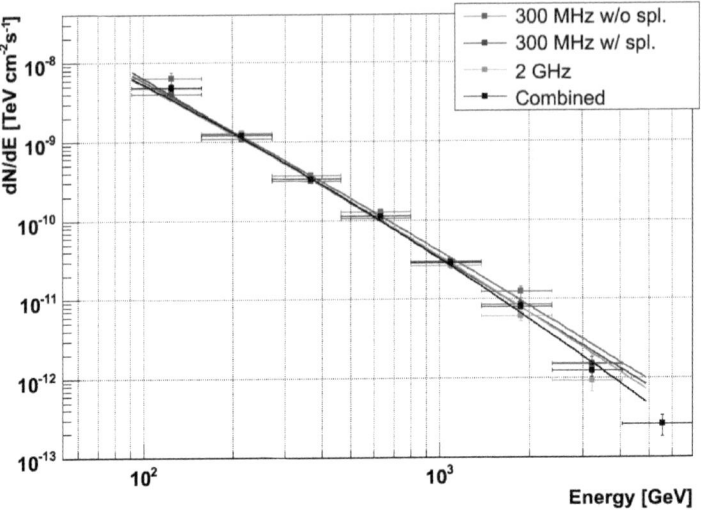

Figure 4.3: Differential energy spectra of the Crab Nebula. Shown are the spectra of the Crab Nebula for the three different time periods in red, blue and green with the corresponding fit curves as well as the combined one in black colour.

Sample	f_0 $[10^{-10}\,(\text{TeV}\,\text{cm}^2\,\text{s})^{-1}]$	a	b
300 MHz w/o spl.	5.75 ± 0.26	-2.15 ± 0.11	-0.11 ± 0.07
300 MHz w/ spl.	5.31 ± 0.19	-2.27 ± 0.09	-0.04 ± 0.05
2 GHz	5.26 ± 0.15	-2.16 ± 0.07	-0.17 ± 0.05
Combined	5.37 ± 0.11	-2.20 ± 0.05	-0.11 ± 0.03

Table 4.2: Fit parameters for the different Crab Nebula data samples. $E_0 = 300\,\text{GeV}$ for all fits. The combined dataset will be used throughout this work to calculate comparative numbers.

5 Observations and analysis results

AGN observations are one of the key science programs of the MAGIC telescope project. Up to 500 hours of observation time is spent each year in order to study known sources in depth or increase the number of detected objects to enhance the statistics of extragalactic VHE sources. The first extragalactic sources having been detected were HBLs. Their jet axis, lying close to the observer's line of sight, beams the γ-radiation in the forward direction and thus enhances the detection probability for these sources in the TeV range. Up to now most of the extragalactic sources belong to the class of HBLs. In the following chapter an observation campaign on systematically selected BL Lac objects, mostly HBLs, spanning four years and including almost 450 hours of observations, will be presented. The key goal of this campaign is to establish BL Lac objects as a VHE source class in astronomy and to investigate their spectral properties, in particular their baseline emission. A list of the observation cycles can be found in table 5.1.

Section 5.1 provides an insight into the set of possible candidate sources for TeV emission among BL Lac objects and the catalogues and source compilations from which possible candidates were selected. Eventually the final set of selection criteria applied as well as the list of sources which remain after the selection is presented. In section 5.2 the observation campaign itself is described and some remarks on the selected sources are given. The results of the analysis are focused on in section 5.3 including a source stacking method for VHE astronomy.

Season	Cycle
$\leq 04/2005$	0^a
$06/2005 - 05/2006$	1
$06/2006 - 04/2007$	2
$05/2007 - 04/2008$	3
$05/2008 - 05/2009$	4

Table 5.1: List of observation cycles of the MAGIC telescope. The observation campaign described in this work was performed in the cycles 1 to 4. [a]The time before cycle 1 including the comissioning of the MAGIC telescope was not declared as a regular cycle with proposed observations, therefore it is marked as cycle 0 here.

5 Observations and analysis results

5.1 Search for TeV candidate BL Lac objects

In order to select promising candidates for γ-ray emission in the VHE regime one has to estimate the flux in this energy range. This can be done in different ways:

- Extrapolating existing HE γ-ray spectra of EGRET or the recently launched FGST into the VHE range. The problem with this method is, that the IC peak of the BL Lac objects often lies in between the HE and the VHE range, making an extrapolation difficult. With the FGST data closing the gap between HE and VHE, this method becomes more attractive.

- Estimating the TeV flux by means of SED model fits to existing lower energy data. This is done in Costamante and Ghisellini (2002) and will be considered in more detail below.

- Assuming quasi-equipartition between the magnetic field and the synchrotron radiation field energy density, the synchrotron and the high energy peaks have the same height in a νF_ν vs ν representation. Thus the energy flux at X-ray energies should be comparable to the one at VHE γ-rays. Stecker et al. (1996) derived scaling laws for an example HBL and used these for predictions of the high energy emission of sources not yet detected in this energy regime. This simple VHE flux estimation is only valid for HBLs and was used for the proposal of the observation campaign described in the next sections.

5.1.1 TeV flux estimation

In Costamante and Ghisellini (2002) the TeV emission of BL Lac objects is estimated using a simple one zone SSC model fit to radio, optical and X-ray data. The best candidates should be objects containing high energy electrons as well as a large population of soft seed photons for the IC process. However, the synchrotron peak frequency may not be too high because the upscattered IC photons then enter the Klein-Nishina region where the cross section for interactions becomes less again. Thus the maximum VHE emission is reached in the Thomson regime for a synchrotron peak frequency of

$$\nu_p \sim 3.8 \cdot 10^{15} B_{\text{Gauss}}^{1/3} \frac{\delta}{1+z} \text{Hz} . \qquad (5.1)$$

As measure for the electron energy the photon flux at $1\,\text{keV}$ is taken. For the soft seed photon population the radio emission at $5\,\text{GHz}$ is an appropriate indicator. The direct measurement in the optical range could be spoiled by the host galaxy emission. Out of 246 BL Lac objects from different catalogues and compilations they find 33 sources fulfilling the selection criteria

5.1 Search for TeV candidate BL Lac objects

being the flux at 5 GHz $F_R > 31.6 mJy$ and the flux at 1 keV $F_X > 1.46 \mu Jy$. Applying an SSC model fit to the data the VHE energy flux level can be estimated.

Another estimation for the VHE flux was described by Stecker et al. (1996) using simple scaling arguments. The Compton peak should get upshifted by a factor $\sim \gamma_p^2$, γ_p being the Lorentz factor of the electrons emitting at the synchrotron peak. As template for all HBLs they used the SED of Mkn 421, one of the strongest HBLs and by now a well established VHE γ-ray source. The upshifting factor found for Mkn 421 is $\sim 10^9$ for the data available at that time and the peak luminosities were comparable, $L_C/L_{\text{sync}} \sim 1$. The derived scaling laws are

$$\frac{\nu_o F_o}{L_{\text{sync}}} \sim \frac{\nu_{\text{GeV}} F_{\text{GeV}}}{L_C} \tag{5.2}$$

$$\frac{\nu_X F_X}{L_{\text{sync}}} \sim \frac{\nu_{\text{TeV}} F_{\text{TeV}}}{L_C}. \tag{5.3}$$

With $L_C/L_{\text{sync}} \sim 1$ it follows

$$\nu_o F_o \sim \nu_{\text{GeV}} F_{\text{GeV}} \tag{5.4}$$

$$\nu_X F_X \sim \nu_{\text{TeV}} F_{\text{TeV}}. \tag{5.5}$$

Based on the unifying scheme and the spectral blazar sequence, the estimation of the VHE flux level for the sources discussed within this work was done using formula 5.5 with the X-ray measurements at 1 keV and the VHE flux predictions at 200 GeV.

5.1.2 Source catalogues and compilations

Additionally to the source list proposed in Costamante and Ghisellini (2002), the following catalogues and compilations were used for selecting promising VHE candidates:

Blazar compilation from Donato et al. (2001)
Donato et al. (2001) have compiled a list of 268 blazars – 136 of them HBLs – observed in the X-ray regime with available X-ray spectra. The data are taken from different X-ray satellite experiments. In the soft X-ray band between 0.1 and 2 keV the majority of the measurements are ROSAT data, whereas the hard X-ray band is covered by observations from the EXOSAT, ASCA and BeppoSAX missions[1]. Archival data for the radio (at 5 GHz) and the optical (at 550 nm) bands as well as the redshift of the sources have been

---
[1] ROSAT: Röntgensatellit, EXOSAT: European X-ray Observatory Satellite, ASCA: Advanced Satellite for Cosmology and Astrophysics, BeppoSAX: Satellite per Astronomia X

5 Observations and analysis results

added to the compilation if available. All sources proposed by Costamante and Ghisellini (2002) for VHE γ-ray emission except three are also contained in the Donato compilation.

Catalogue of BL Lac objects by Nieppola et al. (2006)

Using the Metsähovi radio sample Nieppola et al. (2006) simulated the SEDs of over 300 BL Lac objects by fitting the synchrotron peak with a parabolic function in the νF_ν representation. They also included multifrequency data from the optical as well as X-ray bands and made a classification of the objects according to their synchrotron peak frequency into HBLs ($\nu_p > 10^{16.5}$ Hz), IBLs ($\nu_p \sim 10^{14.5-16.5}$ Hz) and LBLs ($\nu_p < 10^{14.5}$ Hz).

Sedentary multifrequency survey from Giommi et al. (2004)

The sedentary survey from Giommi et al. (2004) comprises a list of sources obtained via cross-correlation of a radio catalogue (NRAO VLA Sky Survey, Condon et al. (1998))) with soft X-ray data (RASSBSC, Voges et al. (1999))[2]. The multifrequency data are completed by optical data from the Palomar and UK Schmidt surveys (Irwin et al., 1994; Yentis et al., 1992). The data base therewith contains 218 entries with $\sim 85\%$ of them BL Lac objects. From this survey one source was selected as described in the next sections.

5.1.3 Selection criteria

From the catalogues and compilations described above promising objects for observations with MAGIC have been chosen based on several selection criteria. The criteria were chosen to assure the best possible detection probability for the MAGIC telescope. In order to enlarge the sample in cycle 2 of regular observations the selection criteria were modified as described in the text. An overview of the selection criteria for the different observation cycles can also be found in table 5.2.

Flux at 1 keV $F_X > 2\mu$Jy

Starting with the argument by Stecker et al. (1996), the flux-to-luminosity ratio in X-rays should be comparable to the one in the VHE range (cf. formula 5.5). The same argument applies for the blazar sequence (Fossati et al., 1998) within the unifying scheme. Assuming comparable peak luminosities in X-rays and in VHE γ-rays and additionally that the measured energy range in both bands lies near the peaks, the energy flux at 1 keV should be comparable to the one at 200 GeV. Thus a minimum flux at 1 keV was chosen to assure a high detection probability within a reasonable observation time at 200 GeV.

---

[2]NRAO VLA: National Radio Astronomy Observatory – Very Large Array, RASSBSC: ROSAT All Sky Survey Bright Source Catalogue

5.1 Search for TeV candidate BL Lac objects

An X-ray flux $F_X = 2\mu Jy$ at 1 keV corresponds to a γ-ray flux of $1.5 \cdot 10^{-11}\,\mathrm{cm^{-2}\,s^{-1}}$ at 200 GeV which is under nominal observation conditions detectable by MAGIC within 5.6 hours. This selection criterion was used for all proposed objects except the ones chosen from Nieppola et al. (2006).

Zenith distance zd at culmination

The lowest possible energy threshold is achieved for air showers coming from the zenith. At higher zenith distances the low energy showers, containing less light, are absorbed in the atmosphere before they reach the telescope due to their longer path. Therefore the zenith distance at the culmination of the source was restricted to $zd_c < 30°$ in combination with a redshift $z < 0.3$ or $z < 0.4$ for cycle 2 of observations (cf. next point). There the criterion was additionally loosened to $zd_c < 45°$ but only for sources with $z < 0.15$ in order to enhance the number of possible source candidates. The zd criterion limits the declination range of source candidates to $-1°14' < \delta < 58°46'$ for the former and $-16°14' < \delta < 73°46'$ for the latter case.

Redshift z

With increasing redshift the absorption of γ-rays by the EBL becomes more and more important. At a redshift $z = 0.3$ the cutoff energy E_{cutoff}, being the energy where the intrinsic source spectrum is attenuated to a factor $1/e$, lies at $\sim 320\,\mathrm{GeV}$ (Kneiske and Dole, 2010). For low zd observations the threshold energy E_{thr} of the MAGIC telescope is $\sim 80 - 120\,\mathrm{GeV}$ which is well below the cutoff energy. Thus the combination of $zd < 30°$ and $z < 0.3$ is justified. A redshift $z = 0.4$ leads to a cutoff energy $E_{\mathrm{cutoff}} \sim 250\,\mathrm{GeV}$. For a higher zd (around 40°), $E_{\mathrm{thr}} \sim 200\,\mathrm{GeV}$, so the redshift was accordingly chosen to $z < 0.15$ where $E_{\mathrm{cutoff}} \sim 630\,\mathrm{GeV}$.

As in the Metsähovi radio sample from Nieppola et al. (2006) a different treatment of the classification is done due to the parabolic fitting of the synchrotron peaks, the selection criteria were different:

Synchrotron peak frequency $\nu_p > 2 \cdot 10^{15}\,\mathrm{Hz}$

Most of the BL Lac objects selected from the compilations and catalogues above are also contained in the radio survey from Nieppola et al. (2006). Allowing also IBLs to enter the observation sample, objects were selected with a minimum synchrotron peak frequency $\nu_p > 2 \cdot 10^{15}\,\mathrm{Hz}$ corresponding to $E(\nu_p) = 8.2\,\mathrm{eV}$.

Flux at synchrotron peak frequency $F(\nu_p) > 10^{-11}\,\mathrm{erg\,cm^{-2}\,s^{-1}}$

According to Stecker et al. (1996), the luminosities of the synchrotron and inverse Compton peaks are assumed to be the same. For an estimation of the VHE flux above 100 GeV

5 Observations and analysis results

a decrease of the inverse Compton peak flux of 40 % is assumed based on the results found in Albert et al. (2008c). The resulting flux at 100 GeV for a minimum peak flux $F(\nu_p) = 10^{-11}\,\mathrm{erg\,cm^{-2}\,s^{-1}}$ is $F_{\mathrm{VHE}} = 6.3 \cdot 10^{-12}\,\mathrm{erg\,cm^{-2}\,s^{-1}}$ including the 40% decline. This flux level ist detectable for MAGIC within a timescale of 15 hours.

For cycle 1 13 objects were chosen from the compilation from Donato et al. (2001). Two of them were already known VHE sources, 1ES 1426+428 and 1ES 2344+514, two of them were discovered in the course of the observation campaign, 1ES 1218+304 (Albert et al., 2006) and 1ES 1011+496 (hint for a signal, discovered unambiguously in a later observation campaign triggered by an optical outburst of the source, Albert et al. (2007a)). The observation campaign for cycle 1 is described in more detail in Albert et al. (2008c) and Meyer (2008). The remaining nine sources were added to the object sample of this work due to an improved analysis chain including the timing information of the air showers. Two sources which are not included here also fulfill the selection criteria: the two well-known VHE emitting HBLs Mkn 421 and Mkn 501.

Cycle	z	zd at culm.	$F_X(1\,\mathrm{keV})$	ν_p	$F(\nu_p)$
1	< 0.30	< 30°	> 2 µJy	–	–
2	< 0.15	< 45°	> 2 µJy	–	–
	< 0.40	< 30°	> 2 µJy	–	–
	< 0.40	< 30°	–	$> 2 \cdot 10^{15}\,\mathrm{Hz}$	$> 10^{-11}\,\mathrm{erg\,cm^{-2}\,s^{-1}}$
3	< 0.30	< 30°	> 2 µJy	–	–

Table 5.2: Criteria for the selection of sources from the different catalogues. The criteria for cycle 2 are disjunctive.

In cycle 2 eight objects were selected including one source proposed for cycle 1 but not observed due to missing observation time slots and one source where the analysis of the data in cycle 1 showed a small hint for a signal. The remaining six objects were selected from Costamante and Ghisellini (2002).

As the selection from the Metsähovi sample was done for cycle 2 observations, the zenith distance at culmination and the redshift were limited to $zd_c < 30°$ and $z < 0.4$, accordingly. From this survey three objects were selected in cycle 2 being the only IBLs of the final observation sample.

For cycle 3 one additional object from Costamante and Ghisellini (2002) was chosen where the redshift was not yet determined by the time of the selection for cycle 2. One object was taken from Giommi et al. (2004) that fulfilled the selection criteria. Table 5.3 shows all selected

sources with the relevant parameters.
During cycle 4 follow up observations were performed for one of the sources.

Source	z	F_X[1]	α_X	$\log(\nu_p)$[2]	$\log(F(\nu_p))$[3]	ref.
1ES 0033+595	0.086[c]	5.66	—	18.93	−10.7	CG[†], N
1ES 0120+340	0.272	4.34	1.93	18.32	−10.6	CG, D[†], G, N
1ES 0229+200[b]	0.1396	2.88	—	19.45	−10.8	CG[†], N
RX J0319.8+1845[b]	0.190	1.76	2.07	16.99	−11.4	D[†], G, N
1ES 0323+022	0.147	3.24	2.46	19.87	−10.2	CG, D[†], G, N
1ES 0414+009[b]	0.287	5.00	2.49	20.74	−10.0	CG, D[†], G, N
1RXS J0441278+150455	0.109	4.74	2.10	—	—	G[†]
1ES 0647+250	0.203[c]	6.01	2.47	18.28	−10.5	CG[†], D, N
1ES 0806+524[b]	0.138	4.91	2.93	16.56	−10.8	CG, D[†], N
1ES 0927+500	0.188	4.00	1.88	21.13	−10.3	D[†], G, N
1ES 1011+496[a]	0.212	2.15	2.49	16.74	−10.8	CG, D[†], N
1ES 1028+511	0.361	4.42	2.50	18.56	−10.9	CG, D[†], G, N
RGB J1117+202	0.1392	7.31	1.90	—	—	CG[†]?, D, G
RXS J1136.5+6737	0.135	3.23	2.39	17.55	−10.9	CG, D[†], G, N
B2 1215+30	0.237	1.59	2.65	15.58	−10.9	CG, D, N[†]
1ES 1218+304[a]	0.182	8.78	2.34	19.14	−10.3	CG, D[†], G, N
2E 1415.6+2557	0.237	3.26	2.25	19.24	−10.5	CG, D[†], G, N
PKS 1424+240[a,b]	0.160[c]	1.37	2.98	15.70	−11.0	D, N[†]
1ES 1426+428[b]	0.129	7.63	2.09	18.55	−10.6	D[†], N
RX J1725.0+1152	0.018[c]	3.60	2.65	15.80	−10.7	CG, D[†], N
1ES 1727+502	0.055	3.36	2.61	17.42	−10.9	CG, D[†], N
1ES 1741+196	0.083	1.92	2.04	17.91	−11.0	CG[†], D, N
B3 2247+381	0.119	0.60	2.51	15.61	−11.0	D, N[†]
1ES 2344+514[a,b]	0.044	4.98	2.18	16.40	−10.8	D[†], N

Table 5.3: List of sources selected for the observation campaign. For each source the redshift, the X-ray flux F_X and spectral index α_X at 1 keV, the synchrotron peak frequency ν_p and the corresponding flux $F(\nu_p)$ as well as the references are given (CG: Costamante and Ghisellini (2002), D: Donato et al. (2001), G: Giommi et al. (2004), N: Nieppola et al. (2006)). The catalogue from where the source was selected is marked with a †.
[a]These sources were detected by MAGIC during the observation campaign as described in section 5.2 or during an independent observation.
[b]These sources were detected by other ground-based VHE γ-ray experiments.
[c]Tentative redshift or lower limit.
[1]in μJy.
[2]in $\log(Hz)$.
[3]in $\log(erg\,cm^{-2}\,s^{-1})$.

5.2 Observation campaign

The observation campaign presented in this work spans over four MAGIC observation cycles between August 2005 and April 2009. In order to achieve comparable results, the very same analysis chain has been used for all sources. Consequently, only observations that were performed in wobble mode are considered throughout this work. This leads to a list of 20 sources

5 Observations and analysis results

previously undetected by MAGIC. The overall observation time of 449.8 hrs in almost four years corresponds to $\sim 22\,\%$ of the extragalactic dark night observations performed by MAGIC. Table 5.4 lists the sources which were chosen for the further analysis as presented in section 5.3 together with the season of observation and the exposure time per source.

Source	Season	$t_{\rm exp}$ [h]	$t_{\rm eff}$ [h]	η_t [%]
1ES 0033+595	08/2006 – 07/2008	16.9	5.2	30.5
1ES 0120+340	08 – 09/2005	22.1	10.7	48.3
1ES 0229+200	08 – 11/2006	14.0	8.0	57.3
RX J0319.8+1845	09/2005 – 01/2006	6.5	4.7	72.8
1ES 0323+022	09 – 12/2005	14.4	11.4	79.3
1ES 0414+009	12/2005 – 01/2006	20.0	18.2	91.2
1RXS J044127.8+150455	10 – 12/2007	32.4	26.9	83.0
1ES 0647+250	02 – 03/2008	32.2	29.2	90.7
1ES 0806+524	10 – 12/2005	21.0	17.5	83.3
1ES 0927+500	12/2005 – 02/2006	24.0	16.7	69.6
1ES 1028+511	03/2007 – 02/2008	42.4	37.1	87.8
RGB J1117+202	01/2007 – 03/2008	18.6	14.9	80.0
RXS J1136.5+6737	02/2007	17.6	14.8	84.3
B2 1215+30	03/2007 – 03/2008	20.7	16.0	77.5
2E 1415.6+2557	02/2007 – 04/2008	53.3	44.4	83.3
PKS 1424+240	05/2006 – 02/2007	21.8	20.0	91.8
RX J1725.0+1152	03/2007 – 04/2009	30.8	27.4	89.0
1ES 1727+502	05/2006 – 05/2007	8.8	6.1	69.0
1ES 1741+196	07/2006 – 04/2007	16.0	11.9	74.2
B3 2247+381	08 – 09/2006	16.3	8.3	51.1
Overall	08/2005 – 04/2009	449.8	349.5	77.7

Table 5.4: Sources selected for the analysis as presented in section 5.3. Listed are the season, the exposure times $t_{\rm exp}$, the effective On time $t_{\rm eff}$ and the efficiency $\eta_t = t_{\rm eff}/t_{\rm exp}$ after quality selection of the data.

5.2.1 Known sources from the selected sample

Four of the sources observed within the observation campaign were detected earlier by other ground-based Cherenkov experiments (Mkn 421, Mkn 501, 1ES 1426+428 and 1ES 2344+514), five sources afterwards (1ES 0229+200, RX J0319.8+1845, 1ES 0414+009, 1ES 0806+524, PKS 1424+240). Five sources from the sample were detected by MAGIC within the cycles 1–4, two of them VHE discoveries (Mkn 421, Mkn 501, 1ES 2344+514; discovery: 1ES 1011+496, 1ES 1218+304). One source was discovered after this campaign by MAGIC (PKS 1424+240, contemporaneously with VERITAS). Thus the selection criteria turned out to be quite effective. The following list describes these sources in more detail ordered by increasing rightascension.

5.2 Observation campaign

1ES 0229+200

Observations by the H.E.S.S. experiment in 2005 and 2006 led to a detection of 1ES 0229+200 with a significance of $6.6\,\sigma$ in 41.8 h observation time (Aharonian et al., 2007). The differential energy spectrum has a relatively hard spectral slope of $2.50\pm0.19_{stat}$ $\pm0.10_{syst}$ between 500 GeV and 15 TeV. The integral flux above 580 GeV is calculated to be $(9.4\pm1.5_{stat}\pm1.9_{syst})\cdot10^{-13}\,\mathrm{cm^{-2}\,s^{-1}}$ corresponding to 1.8 % of the Crab Nebula flux above this energy. The measured flux doesn't show any hint for variability. No signs of flux variability have been found

RX J0319.8+1845

The discovery of RX J0319.8+1845 was recently announced by the VERITAS collaboration (Ong and Fortin, 2009). The observations revealed a signal of 6 standard deviations and the integral flux was determined to be $\sim$ 2% of the Crab Nebula flux above 200 GeV. The observations were performed after the source was detected above 100 MeV with the FGST. The VHE flux lies well below the detectable flux with MAGIC for the observation time given in table 5.4.

1ES 0414+009

Together with the FGST, the H.E.S.S. collaboration recently reported the discovery of this HBL at an integral flux level in the VHE regime corresponding to 0.5 % of the Crab Nebula flux. The signal has a significance of 5 standard deviations in 60 h observation time. (Hofmann and Fegan, 2009). As for RX J0319.8+1845 the reported flux could not have been detected with MAGIC within the observation time given in table 5.4.

1ES 0806+524

VERITAS observed this source between 2006 and 2008 and detected a signal with a significance of $6.3\,\sigma$ in 65 h (Acciari et al., 2009a). The spectral slope of the differential energy spectrum is $3.6\pm1.0_{stat}\pm0.3_{syst}$ between 300 and 700 GeV and the integral flux is calculated to be $(2.2\pm0.5_{stat}\pm0.4_{syst})\cdot10^{-12}\,\mathrm{cm^{-2}\,s^{-1}}$ above 300 GeV. Only little variability was found on monthly time scales.

1ES 1011+496

Within the observation campaign by MAGIC in cycle 1 a hint for a signal from this source could be found at a level of $3.5\,\sigma$ (Albert et al., 2008c). However, in March 2007 it was clearly detected in an observation performed after a trigger due to an optical flux high state as reported by Albert et al. (2007a). The energy spectrum above 120 GeV is very

5 Observations and analysis results

soft and can be described by a single power law fit,

$$\frac{dN}{dE} = (2.0 \pm 0.1) \cdot 10^{-10} \frac{1}{\text{TeV cm}^2 \text{s}} \left(\frac{E}{0.2\,\text{TeV}}\right)^{-4.0\pm0.5}. \tag{5.6}$$

Mkn 421

The first extragalactic VHE source was discovered by the Whipple Collaboration (Punch et al., 1992). Mkn 421 is a well-known and deeply studied HBL. Its vicinity (z=0.03) allows for investigating the intrinsic source spectrum almost without the problem of γ-ray absorption by the EBL. It was detected by several ground-based Cherenkov experiments, e.g. HEGRA (Petry et al., 1996), H.E.S.S. (Aharonian et al., 2005) and MAGIC (Albert et al., 2007c). Like Mkn 501 it is mentioned here because both HBLs fulfill the selection criteria for this observation campaign.

1ES 1218+304

The first extragalactic discovery by MAGIC was 1ES 1218+30.4 (Albert et al., 2006). It was detected even before the cycle 1 observation campaign in January 2005 in a high VHE flux state revealing an energy spectrum compatible with a single power law fit above 140 GeV,

$$\frac{dN}{dE} = (8.1 \pm 2.1) \cdot 10^{-11} \frac{1}{\text{TeV cm}^2 \text{s}} \left(\frac{E}{250\,\text{GeV}}\right)^{-3.0\pm0.4}. \tag{5.7}$$

The observations in 2006 within the campaign described in this work showed a strong hint for a signal (4.6 σ) with a flux level being $\sim 30\%$ lower than in 2005 (Albert et al., 2008c).

PKS 1424+240

Within this observation campaign PKS 1424+240 was not detected (cf. section 5.3). However, after a trigger in April 2009 from FGST measuring the source in a high flux state at high energies, observations performed by VERITAS and MAGIC revealed a $\gtrsim 5\,\sigma$ signal also in the VHE regime (Acciari et al. et al., 2009; Teshima, 2009).

1ES 1426+428

This source is an established VHE γ-ray emitter. It was detected by the Whipple and HEGRA collaborations in 2001 and 1999/2000, respectively (Horan et al., 2002; Aharonian et al., 2003b). Within the MAGIC observation campaign only an upper limit for the integral flux above 190 GeV could be derived to be $F_{>190\text{GeV}} < 1.18 \cdot 10^{-11}\,\text{cm}^{-2}\,\text{s}^{-1}$ (Albert et al., 2008c).

5.2 Observation campaign

Mkn 501

As Mkn 421, Mkn 501 is a close-by HBL at z=0.034. Its VHE emission is thoroughly studied – albeit not yet fully understood – in low as well as high emission states. It was discovered in VHE γ-rays by the Whipple Collaboration (Quinn et al., 1996) and detected by HEGRA (Bradbury et al., 1997) and MAGIC. In 1997 it underwent a major outburst in all wavebands that lasted for weeks. The known strong sources like Mkn 421 and Mkn 501 are monitored continuously with MAGIC. Due to that a large flare could be detected in July 2005 with doubling time scales of the VHE flux of less than four minutes (Albert et al., 2007d) and a delay of the higher energy part compared to the lower energy part in the MAGIC data. Interpreting this delay as a violation of Lorentz invariance, limits could be set to the scale of quantum gravity (Albert et al., 2008b).

1ES 2344+514

Discovered originally by the Whipple collaboration (Catanese et al., 1998), 1ES 2344+514 was detected in cycle 1 by MAGIC in a flux state much lower compared to the one measured by Whipple. The energy spectrum can be described by a single power law fit,

$$\frac{dN}{dE} = (1.2 \pm 0.1_{\text{stat}} \pm 0.5_{\text{syst}}) \cdot 10^{-11} \frac{1}{\text{TeV cm}^2 \text{s}} \left(\frac{E}{500\,\text{GeV}}\right)^{-2.95 \pm 0.12_{\text{stat}} \pm 0.2_{\text{syst}}} \quad (5.8)$$

for energies above 140 GeV (Albert et al., 2007b).

5.2.2 Tentative redshift measurements

For some objects the redshift could not be determined unambiguously. Optical observations of 1ES 0033+59.5 could not resolve the host for a photometric redshift determination. The redshift $z = 0.086$ is stated in the literature to be derived from a private communication with Perlman. Sbarufatti et al. (2005) calculate a lower limit of $z > 0.24$.

For 1ES 0647+250 a tentative redshift $z = 0.203$ has been derived. Falomo and Kotilainen (1999) state a redshift $z > 0.3$ due to the non-detection of the host galaxy. They argue that at $z = 0.203$ the host galaxy would have to be 2 orders of magnitudes fainter than for an average BL Lac object.

The redshift of PKS 1424+240 is under debate. It was included based on a redshift $z = 0.16$ which recently was doubted by Sbarufatti et al. (2005) who give a lower limit of $z > 0.67$ due to a non-detection of the host galaxy.

Similar to PKS 1424+240, the redshift value for RX J1725.0+1152 was revised recently by Sbarufatti et al. (2006) from $z = 0.018$ to a lower limit of $z > 0.17$.

5 Observations and analysis results

Albeit the redshift values for these sources are only tentative, the originally stated values will be used (cf. table 5.3).

5.3 Analysis results

In this section the results of the analysis of the objects observed during the campaign described above will be presented in detail. First the results of the analysis chain (cf. section 5.3) are shown followed by the description and application of the upper limit calculation method by Rolke et al. (2005). Finally a source stacking method for VHE astronomy will be introduced and applied to the data.

5.3.1 Results of the analysis chain

In order to get comparable results the very same analysis chain was applied to all the data of the individual sources as well as to the Crab Nebula samples presented already in section 4.6.

Quality selection

Many different factors affect the quality of the data. Some of them can be corrected within the analysis, some cannot. The main influences are the weather, including different light conditions, and technical reasons like a mispointing due to the bending of the telescope as well as technical problems like broken PMTs for instance. The data which are affected by non-correctable issues have to be excluded from the analysis. The following reasons led to the exclusion of data for the final analysis:

> **Low imaging rate**
>
> In general the observations were supposed to be performed in clear moonless nights. The usual data taking rate after the image cleaning (imaging rate) is $\sim 200\,\text{Hz}$ for good quality data. Depending on the light conditions, e.g. weak moon light or stray light, this rate can be lower without affecting the final γ-rate too much. In addition, with increasing zenith distance the imaging rate is decreasing. Thus all data with an imaging rate above 160-170 Hz were included. With clouds present in the sky, the rate is much lower or fluctuating strongly. Data affected by fluctuating rates were also excluded. A concomitant effect of cloudy weather can be a high relative humidity. Values above $\sim 70\,\%$ indicate bad atmospheric conditions.
>
> An additional weather condition which affects the rates is the so-called calima, a dust

5.3 Analysis results

layer with its origin in the Sahara desert, in a height between 2000 and 5000 m a.s.l. (Albert et al., 2009).

High camera inhomogeneity

The camera inhomogeneity is a technical problem which affects the acceptance of air showers by the trigger logic. Due to some broken PMTs especially low energy air showers which only illuminate a few pixels are not triggering homogeneously different parts of the camera. This leads to a non-homogeneous distribution of accepted events. The inhomogeneity can be characterised by considering the acceptance of the six sectors of the camera separately and calculating the variance among them. For values above 25 % the data are rejected.

The efficiency η_t of the quality selection, being the ratio of the effective On time t_{eff} and the exposure time t_{exp}, can be found in table 5.4.

γ – hadron separation cuts

The cuts for the separation of γ-like and hadronic shower images are the most effective mean for extracting the γ-signal from the data. They are made in the multidimensional image parameter space as explained in chapter 4. The automatic analysis pipeline in the Würzburg data centre provides a set of robust γ – hadron separation cuts in order to handle all data taken since MAGIC went operational. The *area*- and ϑ-cuts are optimised on a data sample of the Crab Nebula by means of a Minuit method. It is made with the goal of minimising the product of significance and logarithm of excess events. The parameters for the *disp* in equations 4.3 and 4.4 are optimised using MC simulated γ-images.

For the analysis presented here the parameter cuts were optimised on a data set of the Crab Nebula trying to cover the whole time and *zd* range of the data. Additionally the same conditions in the quality parameters like the inhomogeneity and the imaging rate should be fulfilled. However, the resulting set of cuts did not show any significant improvement compared to the standard set used in the data centre. Thus the latter one was used throughout the analysis, except the ϑ-cut. The optimisation of this cut on data of the Crab Nebula gets spoiled by the strong signal of this source implying a broader spread of excess events in the ϑ^2-distribution. In case of weak signals like the ones considered here the ϑ-distribution of excess events, if any, is much narrower. Thus the ϑ-cut was chosen to be $\vartheta = 0.14°$ corresponding to a size of the signal region of 2.8 pixels in the camera plane. The choice of ϑ is justified, because the point spread function of a point-like source did not exceed 16.0 mm during the observation campaign,

5 Observations and analysis results

which translates into a diameter of 1.1 camera pixels. The whole set of cut values is listed in table 5.5.

Param.	Value
c_0	1.15136
c_1	0.0681437
c_2	2.62932
c_3	1.51279
c_4	0.0507821
c_5	-0.07
c_6	0.5
c_7	7.2
c_8	0.215468
c_9	5.63973
c_{10}	0.0836169
c_{11}	0.14
$c_9^\star$	5.49973
$c_{11}^\star$	0.215

Table 5.5: γ – hadron separation cut values. The two values changed for the energy estimation are marked with a $\star$.

The recorded data sequences and the data used for the analysis after quality selection for all objects including the Crab Nebula are listed in appendix A.
After applying the γ – hadron separation cuts to the data one gets the number of events in the signal region, being the region inside the ϑ-cut, and the corresponding excess and background events. From these numbers the significance of the signal is calculated using formula 17 from Li and Ma (1983). The measured values for all sources are listed in table 5.6. The corresponding ϑ^2-plots can be found in appendix B. As an example, the ϑ^2-distribution after γ – hadron separation for 1ES 1028+511 is shown in figure 5.1. There the events inside the source signal region (black crosses) and the three background regions (grey shaded area) are plotted. The dashed line represents the ϑ^2-cut below which the excess events are calculated.

Energy estimation

For the γ – hadron separation the cuts are optimised to obtain the best significance without loosing too many excess events. For the energy estimation this approach is not optimal because the energy threshold E_{thr} of the analysis is increased artificially. The energy threshold of the separation cuts described above is $\sim 200\,\text{GeV}$. The optimisation routine is rejecting the low energy events because they cannot be separated well as can be seen in figure 5.2. There the *area* vs. *size* distributions of MC simulated γ-events and real background data after image cleaning merge at energies below ~ 200 photo electrons corresponding to $\sim 150\text{-}200\,\text{GeV}$ depending

5.3 Analysis results

Source	t_{eff} [h]	N_{sig}	N_{bck}	N_{exc}	σ	E_{thr} [GeV]	UL [f.u.]	UL [c.u.]
1ES 0033+595	5.2	391	331.0	60.0	2.8	165	2.42	8.55
1ES 0120+340	10.7	458	437.3	20.7	0.9	121	3.13	7.46
1ES 0229+200	8.0	627	572.0	55.0	2.0	121	5.09	14.72
RX J0319.8+1845	4.7	285	309.3	-24.3	-1.2	121	2.66	7.70
1ES 0323+022	11.4	706	751.3	-45.3	-1.5	165	1.68	5.92
1ES 0414+009	18.2	1092	1020.7	71.3	1.9	165	1.92	6.78
1RXS J044127.8+150455	26.9	1844	1825.7	18.3	0.4	121	1.21	2.89
1ES 0647+250	29.2	1862	1797.7	64.3	1.3	121	1.59	3.80
1ES 0806+524	17.5	769	752.0	17.0	0.5	141	2.16	7.61
1ES 0927+500	16.7	731	702.7	28.3	0.9	141	1.70	4.92
1ES 1028+511	37.1	2378	2312.3	65.7	1.2	141	1.01	2.92
RGB J1117+202	14.9	830	804.3	25.7	0.8	141	2.00	4.78
RXS J1136.5+6737	14.8	977	954.3	22.7	0.6	226	0.85	4.45
B2 1215+30	16.0	1114	995.0	119.0	3.2	121	3.51	8.38
2E 1415.6+2557	44.4	2682	2686.7	-4.7	-0.1	121	1.65	3.94
PKS 1424+240	20.0	1262	1210.3	51.7	2.0	121	3.10	7.41
RX J1725.0+1152	27.4	1764	1707.7	56.3	1.2	141	1.28	3.69
1ES 1727+502	6.1	333	302.0	31.0	1.5	141	3.58	10.36
1ES 1741+196	11.9	830	731.3	98.7	3.1	121	3.61	8.62
B3 2247+381	8.3	512	490.3	21.7	0.8	141	1.64	4.73
Overall	349.5	21447	20694.0	753.0	4.5	150	0.45*	1.40*

Table 5.6: Analysis results for the X-ray selected BL Lac objects. Listed are the effective on time t_{eff}, the number of signal, background and excess events N_{sig}, N_{bck}, N_{exc} as well as the significance σ of the signal. The energy thresholds E_{thr} of the analysis are achieved by loosening some of the γ – hadron separation cuts, namely the ϑ- and the area-cuts. Additionally the integral upper limits (UL) in flux units (f.u.) of 10^{-11} photons cm^{-2} s^{-1} and in % Crab units (c.u.) above E_{thr} are given. The overall numbers are taken from the stacking analysis. The corresponding energy threshold is fixed to 150 GeV.
*Measured flux above 150 GeV.

mainly on the zenith distance of the observation. In order to obtain a lower energy threshold some of the separation cuts have to be loosened. The most efficient separation is achieved by the ϑ- and area-cut, which are opened by changing parameters c_9 and c_{11}, respectively. By loosening these cuts, more excess events can be included in the energy estimation on the expense of the sensitivity because disproportionately more background events have to be handled. The gain in excess events at lower energies leads to a lower energy threshold of the analysis. The loosened area-cut is also shown in figure 5.2. As the energy threshold is depending on the source spectrum, MC simulations with an input spectral index of $\alpha = 3.0$ are used. This value accounts for softer spectra measured for BL Lac objects compared to the Crab Nebula spectrum. The values of E_{thr} obtained for the different objects are listed in table 5.6, too.

The loose cut values for ϑ and area were optimised (maximising the value significance times log(excess events)) on Crab Nebula data which were divided into subsamples depending on the zenith distance and the different DAQ systems (300 MHz vs. 2 GHz). However, neither of them

5 Observations and analysis results

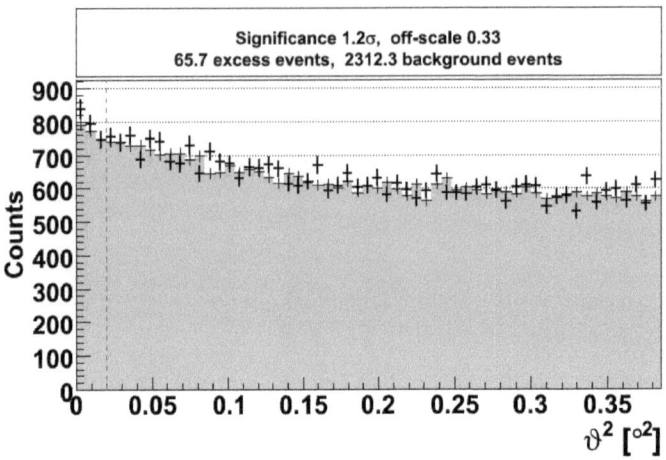

Figure 5.1: ϑ^2-plot for 1ES 1028+511 after detection cuts. Black crosses mark the events from the source signal region, the grey shaded area mark events from the three background regions, scaled with 1/3. The dashed vertical line represents the ϑ^2-cut.

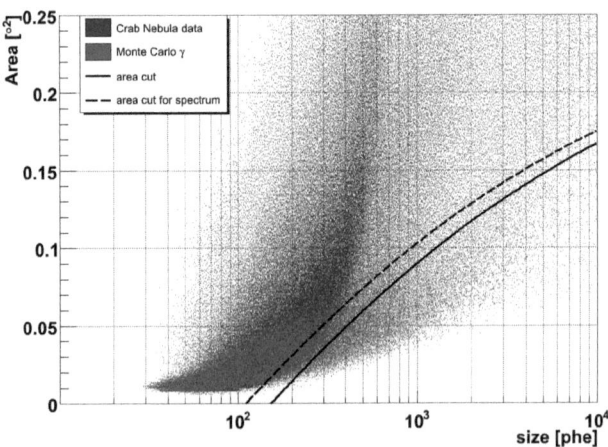

Figure 5.2: Comparison of the *area* vs *size* distributions of data (blue) and MC simulations (red). In addition the applied *area*-cuts are displayed as black (standard) and dashed black (loosened) lines.

5.3 Analysis results

had a large effect on the cut values leaving one value each for all subsamples. The cut values are also listed and marked with a * in table 5.5.

As none of the objects showed a significant signal, the energy estimation was only used for determining the energy thresholds. Individual energy spectra could not be calculated for any of the objects.

Variability

One of the most important conditions for postulating a steady-state emission is the lack of significant variability in the lightcurves of the sources. Throughout the observation campaign none of the sources showed any active or variable behaviour. The lightcurves for the individual sources can be found in appendix C. As an example figure 5.3 displays the excess rates on diurnal basis for RXS J1136.5+6737. All measured excess values are compatible with the mean value. Deviations in the background values presumably arise from influences which were not mirrored in the quality parameters, for instance an error inside the DAQ leading to a larger dead time. As the measurements underlie statistical fluctuations, also negative rates are possible, in particular with low level signals. Variability is defined here as a deviation from the mean value of more than three standard deviations.

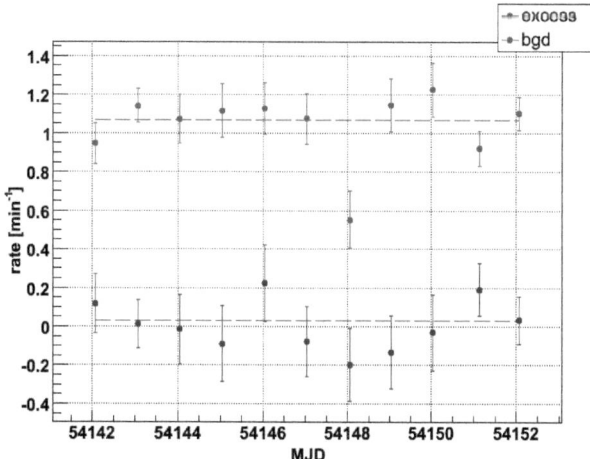

Figure 5.3: Lightcurve based on excess (blue) and background (red) event rates for RXS J1136.5+6737. The dashed lines represent mean values.

5 Observations and analysis results

5.3.2 Upper limit calculation

None of the objects showed a significant signal in the analysis. Therefore upper limits (ULs) on the integral flux above the individual energy thresholds have been calculated.

Method

For the UL calculation, the method from Rolke et al. (2005) has been used. It computes a confidence interval on the excess rate of a Poisson distributed signal by using the profile likelihood method with a frequentist approach. It is implemented in the ROOT framework as the class TRolke allowing for different background and efficiency distributions. The background distribution can be either Gaussian, Poisson or known. An additional efficiency can be included which has either a Gaussian, Binomial or known distribution. The efficiency in case of the analysis presented here can be interpreted as a systematic error of the analysis. For each combination of the given background and efficiency distributions a different model is inferred. Due to the discrete background distribution, a Poisson model is assumed for the data analysed here. As for the calculation of the significance usually no systematic errors are taken into account, the calculation of the confidence intervals should also disregard any systematics. Therefore the efficiency is known and set to 1. With these boundary conditions model 4 in TRolke is used, requiring the number of observed background and signal events, N_{bck} and N_{sig}. Additionally the ratio of background to signal statistics and the confidence level have to be given as input. The confidence level for the ULs calculated here shall be 99.7% corresponding to 3 standard deviations.

Application

Table 5.6 lists all integral ULs above the corresponding energy threshold calculated with the Rolke method 4 for the individual objects. Additionally the integral ULs above the threshold in units of the Crab Nebula flux are given.

5.3.3 Significance distribution

It is striking that all measured significance values shown in table 5.6 except three are positive. In case of the absence of any signal the values should be 0. Due to statistical fluctuations in the background measurement the distribution of the individual signals should be Gaussian with a mean value $m_G = 0$ and a standard deviation $\sigma_G = 1$. In figure 5.4 the distribution of the significances calculated in the analysis is plotted. In fact the shift to positive values is clearly visible. A Gaussian fit to the data yields a mean value $m_{\text{BLLac}} = 1.41 \pm 0.29$ and a standard deviation $\sigma_{\rm BLLac} = 0.98 \pm 0.42$. The positive shift in the mean value and the lack of significant variability implies that the X-ray selection is effective in finding a promising sample for VHE γ-ray steady-state emission. As none of the sources was detected strongly, i.e. with a significance of > 4 standard deviations, the stacking of the individual signals to a cumulative sample will be performed as described in the next section.

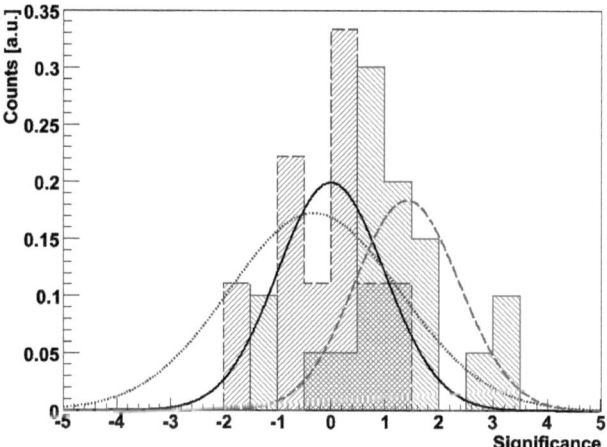

Figure 5.4: Distribution of significances of the blazar sample (red histogram, hatched up left to low right) and a crosscheck sample (cf. section 5.3.5, blue histogram, hatched low left to up right), normalised to unity. The red dashed and blue dotted curves represent Gaussian fits to the corresponding data. The black solid curve shows a Gaussian distribution with a mean $m = 0$ and a standard deviation $\sigma = 1$.

5.3.4 Source stacking

Source stacking is a commonly used approach in astronomy applied to signals below the actual sensitivity of the instrument. The stacking of these signals can allow the cumulative result to exceed the detection limit. Conclusions are then only possible for the whole population of stacked signals, information on the individual objects get lost due to the statistical characteristics of the method.

5 Observations and analysis results

Method

As there are different approaches for stacking signals, the method used here for VHE γ-ray data taken with the MAGIC telescope will be explained in detail. For each source the signal, background and excess events after γ – hadron separation are calculated and illustrated by the ϑ^2-plots. In order to obtain the stacked ϑ^2-distribution the individual numbers for each object, N_{sig}, N_{bck} and N_{exc}, just have to be added up binwise to achieve the final stacked distribution. This is only possible if the ϑ^2-distributions are binned in exactly the same way. But this condition is fulfilled by the usage of the same analysis chain for all sources.

In case of the energy spectrum some more parameters have to be taken into account. The differential energy spectrum can be calculated binwise using formula 4.21:

$$\frac{\mathrm{d}N_i}{\mathrm{d}E} = \frac{N_{\text{exc},i} \cdot a_i}{t_{\text{eff}} \cdot A_{\text{eff},i} \cdot \Delta E_i} \tag{5.9}$$

Therefore one has to include the individual spill over factors a_i, the effective collection areas $A_{\text{eff},i}$, the overall effective On time t_{eff} and the energy bin widths ΔE_i for each bin i. The bin widths again need to be the same for all sources. For the stacking the individual numbers $N_{\text{exc},i}$ and t_{eff} of each source have to be added up. The spill over factor and the effective collection area for each bin is calculated as mean of all corresponding values weighted with the effective On time:

$$\langle a_i \rangle = \frac{\sum_n a_{i,n} \cdot t_{\text{eff},n}}{\sum_n t_{\text{eff},n}} \tag{5.10}$$

$$\langle A_{\text{eff},i} \rangle = \frac{\sum_n A_{\text{eff},i,n} \cdot t_{\text{eff},n}}{\sum_n t_{\text{eff},n}} \tag{5.11}$$

The differential energy spectrum for each stacked bin is then calculated as

$$\frac{\mathrm{d}N_i}{\mathrm{d}E} = \frac{\sum_n N_{\text{exc},i,n} \cdot \langle a_i \rangle}{\sum_n t_{\text{eff},n} \cdot \langle A_{\text{eff},i} \rangle \cdot \Delta E_i} \tag{5.12}$$

with n being the number of objects to be stacked.

Application

Applying this method to the data leads to the ϑ^2-plot shown in figure 5.5. The stacking of the individual results yields a signal with 753 excess above 20694 background events corresponding to a significance of $4.5\,\sigma$ in 349.5 h. A spectrum can be extracted from the excess events according to equation 5.12, which was not possible for the individual objects. The differential

5.3 Analysis results

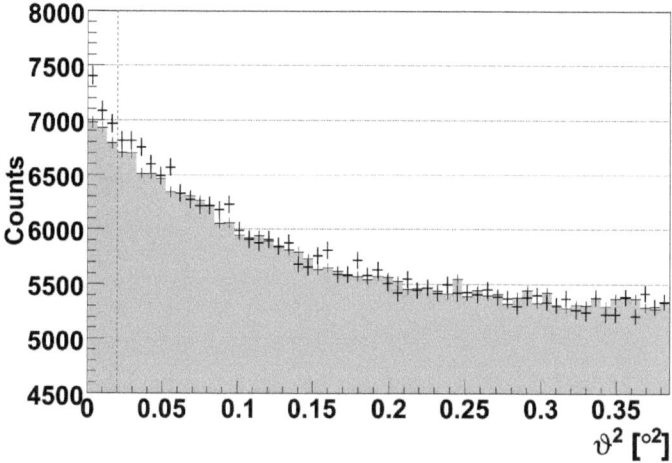

Figure 5.5: ϑ^2-distribution of the stacked blazar sample. The result shows a clear excess with a significance of 4.5 σ (753 excess and 20694 background events).

energy spectrum is shown in figure 5.6 and can be fitted with a power law above 150 GeV lying above most of the individual energy thresholds:

$$\frac{dN}{dE} = (2.60 \pm 0.76) \cdot 10^{-11} \frac{1}{\text{TeV cm}^2 \text{ s}} \cdot \left(\frac{E}{200 \,\text{GeV}}\right)^{-3.15 \pm 0.57} \quad (5.13)$$

The differential flux at 200 GeV is equal to 1.81 % of the Crab Nebula flux. The integral flux above 150 GeV is $F = 4.48 \cdot 10^{-12}$ ph cm^{-2} s^{-1} corresponding to 1.40 % of the Crab Nebula flux.

In case of a homogeneous distribution of the individual source signals, the cumulative excess N_{exc} and the significance σ should increase with t_{eff} and $\sqrt{t_{\text{eff}}}$, respectively. In order to test for such a behaviour, both parameters have been plotted in figure 5.7, ordering the sources in terms of rightascension (as e.g. in table 5.6). Although one cannot expect that the excess is distributed homogeneously on each object, the left panel of figure 5.7 shows a linear increase of the cumulative excess within the errors.

The linear fit yields

$$N_{\text{exc}} = (2.00 \pm 0.27) \frac{1}{\text{h}} t_{\text{eff}} \quad . \quad (5.14)$$

79

5 Observations and analysis results

In the bottom panel of figure 5.7 the distribution of the cumulative significance vs t_{eff} can be fitted with a $\sqrt{t_{\text{eff}}}$ function resulting in

$$\sigma = (0.22 \pm 0.03)\frac{1}{\sqrt{h}}\sqrt{t_{\text{eff}}} \quad . \tag{5.15}$$

Both plots shows that the excess is in general distributed over all objects, demonstrating that all sources contribute to the stacked signal and showing that it is not caused by a few individual strong signals. However, there are deviations from a homogeneous distribution of the signal and the scatter of the values in particular at low t_{eff} is caused by this inhomogeneity.

5.3.5 Crosscheck analysis

In order to exclude potential systematic biases caused by the analysis procedure as the origin of the stacked excess, a crosscheck was done. Again the very same analysis chain as for the Crab Nebula and the BL Lac objects was used.

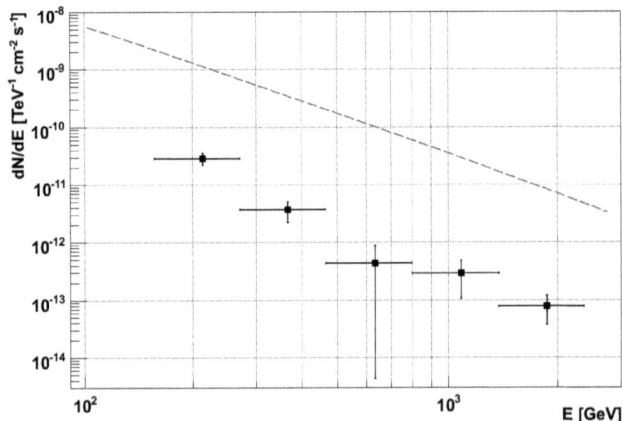

Figure 5.6: Differential VHE spectrum obtained from the stacked data. The grey dashed curve represents the Crab Nebula spectrum (cf. section 4.6) for comparison.

5.3 Analysis results

Sample

The data chosen for the sample are observations where no VHE γ-ray detection is to be expected. In table 5.7 the pointings with the corresponding effective On time and zd range are

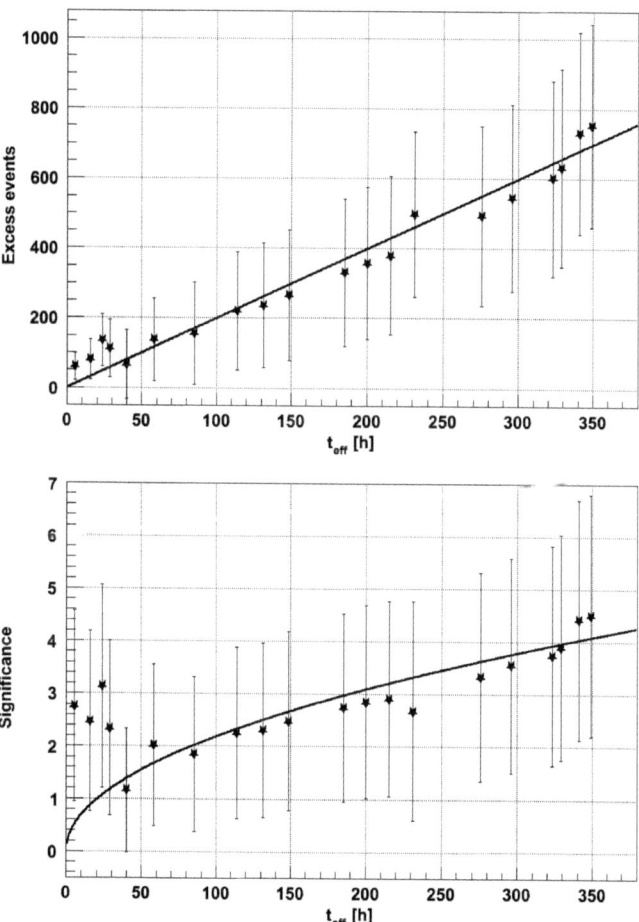

Figure 5.7: Upper panel: Cumulative excess events N_{exc} versus effective On time t_{eff}. The distribution grows linearly with 2.00 N_{exc} per hour (solid line). Lower panel: Distribution of the cumulative significance σ vs t_{eff}. In particular for low values of t_{eff} the distribution shows deviations from a homogeneous distribution. The significance grows by 0.22 per square root hour (solid curve).

5 Observations and analysis results

listed. In most cases the pointings are Off sources for corresponding On observations. In two cases an On pointing was taken of a source from the dark matter search programme of MAGIC, the dwarf spheroidal galaxies Willman 1 and Draco. In Albert et al. (2008d) and Aliu et al. (2009) it is shown that these objects do not exhibit any significant VHE γ-ray emission and upper limits on the fluxes were calculated. All crosscheck datasets were analysed as Wobble mode data. That is the On (or Off) dataset was split into two equally sized datasets. Then one was analysed with respect to a source position 0.4° off centre in positive RA direction, the other one 0.4° off centre in negative RA direction. The combination of both in the end is a quasi-Wobble mode observation of this pointing. Therefore the same analysis chain as for the blazar sample can be used. In addition, in case of a weak γ-ray excess from the original source position this would be outside the new source region defined by the ϑ^2-cut.

Sample	t_{eff} [h]	zd [°]	N_{sig}	N_{bck}	N_{exc}	σ
1	5.4	34 – 43	334	335.3	−1.3	−0.1
2	3.1	6 – 29	112	107.7	4.3	0.4
3	1.9	37 – 47	274	255.0	19.0	1.0
4	3.3	49 – 56	125	149.7	−24.7	−1.8
5	2.8	11 – 27	130	139.7	−9.7	−0.7
6	1.3	28 – 37	78	76.0	2.0	0.2
7	7.3	29 – 36	336	356.7	−20.7	−1.0
8	17.9	22 – 38	1048	1041.0	7.0	0.2
9	9.3	22 – 26	566	548.0	18.0	0.7
Overall	52.3	6 – 56	3003	3009.0	−6.0	−0.1

Table 5.7: Analysis results for the crosscheck sample. Listed are the effective on time t_{eff}, the number of signal, background and excess events N_{sig}, N_{bck}, N_{exc} as well as the significance σ of the signal.

Result

The results of the individual analyses of the crosscheck sample are also listed in table 5.7. The distribution of significances as can be seen in figure 5.4 shows a clear clustering around 0 which is expected for the lack of any γ-ray signal from the individual sources. A Gaussian distribution fitted to the data yields a mean value $m_{\text{X−check}} = -0.31 \pm 1.03$ and a standard deviation $\sigma_{\text{X−check}} = 1.60 \pm 1.55$, fully compatible with the expectations. More details on the data sequences selected for the crosscheck can be found in appendix A. The stacked ϑ^2-distribution shows no excess at small values of ϑ^2 (cf. figure 5.8).

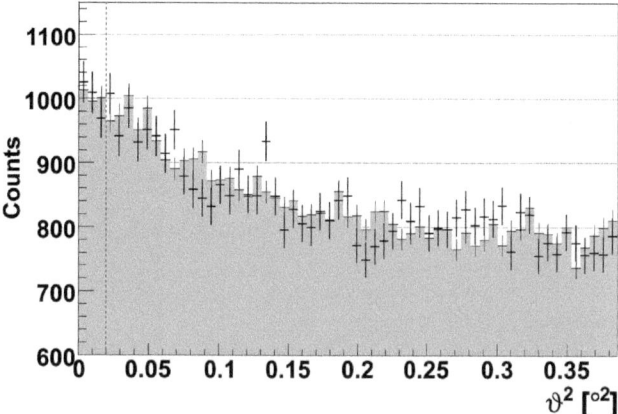

Figure 5.8: Stacked ϑ^2-distribution of the crosscheck sample. As expected, the distribution does not show any excess at small values of ϑ^2 with $-0.1\,\sigma$ (-6 excess and 3009 background events).

5 Observations and analysis results

6 Steady state emission of blazars

The detection of a signal in ~ 350 h, free of any variability, by means of stacking the undetected blazars indicates that a fundamental steady process inside these sources is at work, converting a constant amount of energy into VHE γ-rays. In order to understand the broad-band spectral features of this ensemble the mean flux values in different energy bands will be used in the following to produce a representative steady-state SED (cf. section 6.1). Several sources have been already detected in the VHE regime in a low emission state. The stacked blazar spectrum and SED will be compared to these sources in section 6.2.

6.1 Spectral characteristics

For the mean SED multi-wavelength data from the radio up to the X-ray band have been used. Archival data are available in the radio band at 1.4 GHz and 5 GHz as well as at an X-ray energy of 1 keV (cf. section 5.1.3). In the optical R-band (640 nm) contemporaneous data have been taken with the KVA[1] telescope in La Palma. In case of multiple flux values the arithmetic mean was taken. The mean value $\langle X \rangle$ of the measurements of all objects together is calculated by averaging the individual values X_n weighted with their effective On time $t_{\text{eff},n}$:

$$\langle X \rangle = \frac{\sum_n X_n \cdot t_{\text{eff},n}}{\sum_n t_{\text{eff},n}} \tag{6.1}$$

In order to compare the SED in the jet frame of the sources, the VHE spectrum is corrected for EBL absorption and the luminosities νL_ν in the different energy bands are calculated using the following cosmological parameters: $H_0 = 73 \,\text{km}\,\text{s}^{-1}\,\text{Mpc}^{-1}$, $\Omega_m = 0.27$, $\Omega_\Lambda = 0.73$.

6.1.1 multi-wavelength data

In the radio band at 1.4 GHz the VLA observatory provides a catalog consisting of almost $2 \cdot 10^6$ radio sources above a flux of $\sim 2.5\,\text{mJy}$ (Condon et al., 1998) including all objects from the

---
[1]Kungliga Vetenskaplika Academy

6 Steady state emission of blazars

BL Lac object sample. The values at 5 GHz were taken from the compilations of Donato et al. (2001) and Costamante and Ghisellini (2002) (cf. references therein). One object is missing as it is only contained in the sedentary survey catalogue: 1RXS J044127.8+150455. The radio data are listed in table 6.1. Following the argument of Fossati et al. (1998), the radio luminosity is associated with the behaviour of the sources at higher energies, for example the synchrotron peak flux or energy. Furthermore the source class is determined from the radio luminosity. In table 6.1 also the redshift and the corresponding radio luminosities of the objects are given. The radio data are archival data, as no contemporaneous data could be retrieved during the VHE observations.

In turn, the optical observations are performed at the same time as the VHE observations with the KVA optical telescope in La Palma. For 16 objects the optical flux at 640 nm was measured during the MAGIC observations. In each case the average flux of the measurements is calculated. In the optical band usually the host galaxy spoils the emission of the core of the AGN. Thus the host galaxy flux has to be subtracted from the measurement (Scarpa et al., 2000a,b; Nilsson et al., 2003, 2007). The flux values are also listed in table 6.1.

The X-ray fluxes and photon spectral indices at 1 keV are available for all objects of the sample and are listed in table 5.3. They are adopted in table 6.1. Only for 1ES 0033+595 and 1ES 0229+200 the spectral index is missing, therefore the mean value of all other objects is taken. The knowledge of the spectral behaviour of the sources at 1 keV provides valuable information about the synchrotron peak in the SED.

6.1 Spectral characteristics

Source	$F_{r(1.4\,GHz)}$ [Jy]	$F_{r(5.0\,GHz)}$ [Jy]	$F_{o(640\,nm)}$ [mJy]	$F_{o(host)}$ [mJy]	$F_{o(nucleus)}$ [mJy]	$F_{X(1\,keV)}$ [µJy]	α_X
1ES 0033+595	0.147	0.066	0.49	0.22	0.27	5.66	—
1ES 0120+34.0	0.046	0.040	0.28	0.18	0.10	4.34	1.93
1ES 0229+200	0.082	0.049	1.12	0.98	0.14	2.88	—
RX J0319.8+1845	0.023	0.017	0.27	0.17	0.10	1.76	2.07
1ES 0323+02.2	0.068	0.042	0.77	0.38	0.39	3.24	2.46
1ES 0414+00.9	0.121	0.070	—	—	—	5.00	2.49
1RXS J044127.8+150455	0.014	—	—	—	—	4.74	2.10
1ES 0647+25.0	0.096	0.075	1.06	0^a	1.06	6.01	2.47
1ES 0806+52.4	0.182	0.175	2.40	0.69	1.71	4.91	2.93
1ES 0927+50.0	0.022	0.015	—	—	—	4.00	1.88
1ES 1028+511	0.038	0.044	0.83	0.10	0.73	4.42	2.50
RGB J1117+202	0.103	0.074	1.34	0.66	0.68	7.31	1.90
RXS J1136.5+6737	0.046	0.044	1.05	0.86	0.19	3.23	2.39
B2 1215+30	0.572	0.445	4.40	1.00	3.40	1.59	2.65
2E 1415.6+2557	0.090	0.047	0.95	0.53	0.42	3.26	2.25
PKS 1424+240	0.430	0.316	5.54	0^a	5.54	1.37	2.98
RX J1725.0+1152	0.120	0.091	—	—	—	3.60	2.65
1ES 1727+502	0.201	0.159	2.20	1.23	0.97	3.36	2.61
1ES 1741+196	0.301	0.281	3.22	2.20	1.02	1.92	2.04
B3 2247+381	0.103	0.119	1.32	0.69	0.63	0.60	2.51
Mean value	0.154	0.109			1.22	3.86	2.44

Table 6.1: List of multi-wavelength data used to calculate a broad-band SED. In case of multiple flux values at one energy, the values have been averaged.
^{a}No host galaxy detected.

6 Steady state emission of blazars

6.1.2 EBL correction

As denoted in section 2.2.2, the VHE γ-rays interact with photons from the infrared background. This leads to a deformation of the measured spectrum with respect to the emitted one in dependence of the emitted photon energy. The correction of that effect is done based on the so-called Kneiske-Low EBL model (Kneiske and Dole, 2010) which represents a lower limit on the flux of the EBL. It thus provides a lower limit on the correction of the VHE γ-ray spectrum.

The EBL correction is usually been done on an individual spectrum of a source with known redshift. In case of the stacked energy spectrum an ensemble of objects with different redshifts $0.018 < z < 0.361$ has been used. The spectrum is therefore deabsorbed assuming a mean redshift $z_{\mathrm{mean}} = 0.19$ derived from the mean luminosity distance of the sources calculated as described in section 6.1, $d_L = 886\,\mathrm{Mpc}$. The optical depths τ and deabsorbed flux values for the stacked spectrum can be found in table 6.2. The VHE luminosity values νL_ν for the SED in the figures 6.3 and 6.5 are deabsorbed in the same way.

Energy [GeV]	$\mathrm{d}N/\mathrm{d}E(z=0)$ [(TeV cm^2 s)$^{-1}$]	τ	e^τ	$\mathrm{d}N/\mathrm{d}E(z=0.19)$ [(TeV cm^2 s)$^{-1}$]
214	$2.87\cdot 10^{-11}$	0.25	1.29	$3.69\cdot 10^{-11}$
368	$3.63\cdot 10^{-12}$	0.63	1.87	$6.79\cdot 10^{-12}$
633	$4.36\cdot 10^{-13}$	1.20	3.33	$1.45\cdot 10^{-12}$
1089	$2.92\cdot 10^{-13}$	1.81	6.13	$1.79\cdot 10^{-12}$
1874	$7.74\cdot 10^{-14}$	2.21	9.11	$7.05\cdot 10^{-13}$

Table 6.2: Measured fluxes $F(z=0)$, optical depths τ (taken from Kneiske and Dole, 2010), absorption coefficients e^τ and the deabsorbed flux values for the differential energy spectrum.

The deabsorption leads to a harder differential energy spectrum in the jet frame. Figure 6.1 shows the measured differential energy spectrum (cf. section 5.3) in black and the deabsorbed spectrum in red colour. For comparison the Crab Nebula spectrum is included as dashed grey line (cf. section 4.6). The deabsorbed spectrum can be fitted with a power law according to equation 5.13 resulting in

$$\frac{\mathrm{d}N}{\mathrm{d}E} = (3.47 \pm 1.69) \cdot 10^{-11} \frac{1}{\mathrm{TeV\,cm^2\,s}} \cdot \left(\frac{E}{200\,\mathrm{GeV}}\right)^{-1.93\pm 0.39}. \tag{6.2}$$

6.1 Spectral characteristics

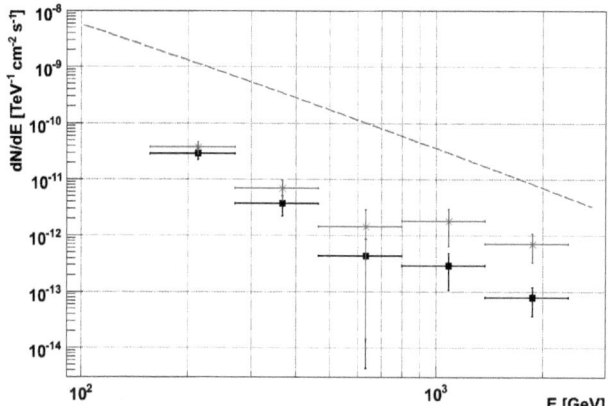

Figure 6.1: Differential energy spectrum for the stacked BL Lac object sample. Shown are the measured spectrum in black and the deabsorbed spectrum in red colour. For comparison the measured Crab Nebula spectrum is plotted as dashed grey line.

6.1.3 Result

The spectral energy distribution provides valuable information about the spectral behaviour of an object and can constrain models of particle acceleration in the source. In case of the stacked BL Lac object sample general statements about the mean SED of blazars in the baseline emission state can be met.

Broad-band spectral indices

Broad-band spectral indices α_{i-j} provide valuable information about the spectral behaviour of sources. They are defined as the quotient of the ratio of the fluxes – given in units of $\text{cm}^{-2}\,\text{s}^{-1}$ – and the ratio of the frequencies in two different energy bands (Ledden and Odell, 1985):

$$\alpha_{i-j} = -\frac{\log(F_i/F_j)}{\log(\nu_i/\nu_j)}, \quad \nu_i < \nu_j \qquad (6.3)$$

For the examination of the SEDs of the individual objects and the mean SED the broad-band spectral indices $\alpha_{X-\gamma}$ between the X-ray and VHE γ-ray regime are calculated using the gathered spectral information at 1 keV and 200 GeV.

89

6 Steady state emission of blazars

According to Stecker et al. (1996) and Fossati et al. (1998) the X-ray energy flux $\nu_X F_X$ should be comparable to the VHE γ-ray flux $\nu_\gamma F_\gamma$. This assumption led to the selection criteria as described in section 5.1.3. In case of equipartition of the magnetic field and synchrotron radiation field energy density, the synchrotron and inverse Compton peak in the SSC model have the same height implying a broad-band spectral index $\alpha_{X-\gamma} = 1.00$. Albert et al. (2008c) find that the maximum value for $\alpha_{X-\gamma}$ equals 1.12 for detected HBLs while the minimum value does is not larger than 0.97. The lower limits found there only exclude values below $\alpha_{X-\gamma} = 0.94$. Figure 6.2 is adopted from Albert et al. (2008c) and shows the lower limits presented there and as determined in chapter 5. For six of the objects stricter lower limits could be determined due to an improvement in the analysis by taking into account the timing information of the recorded air showers. One can clearly see that a broad-band spectral index $\alpha_{X-\gamma} > 1.00$ is favored within the source class of HBLs implying a higher energy output in the X-ray than in the VHE γ-ray regime. From the result of the stacking analysis one can infer that X-ray bright BL Lac objects emit VHE γ-rays in a steady-state mode. Furthermore these objects exhibit in general broad-band spectral indices $\alpha_{X-\gamma}$ larger than unity implying a higher energy output in the X-ray than in the VHE range. This observation goes in line with the spectral blazar sequence of Fossati et al. (1998), showing a weaker emission in the inverse Compton than in the synchrotron regime for HBLs.

The scattering of the values can be mainly traced back to the different observation times for the objects, as deeper observations leave stricter upper limits or even a detection. Furthermore the class of HBLs may have the same spectral characteristics, but no BL Lac objects look alike. Varying Compton peak energies and fluxes as well as different evolutionary states within the source class can lead to a scattering in the measured VHE flux values. In the jet frame this argument becomes even stronger as the γ-ray absorption within the EBL induces a natural scattering to the measured flux values due to the different redshifts. Additionally, the individual spectral energy distributions are shifted in luminosity depending on the distance of the objects.

The broad-band spectral index $\alpha_{X-\gamma}$ of the stacked blazar sample provides an estimate of the mean ratio of the energy outputs in the two energy bands and concludes nicely the spectral behaviour of the HBL/IBL source class as discussed above. For the stacked signal it is determined to $\alpha_{X-\gamma} = 1.09$ (cf. figure 6.2) implying a 5.59 times higher flux level at 1 keV than at 200 GeV. This underlines the result found with the lower limits of the individual objects that the VHE γ-ray energy output is weaker than the one in X-rays. The values for the individual

6.1 Spectral characteristics

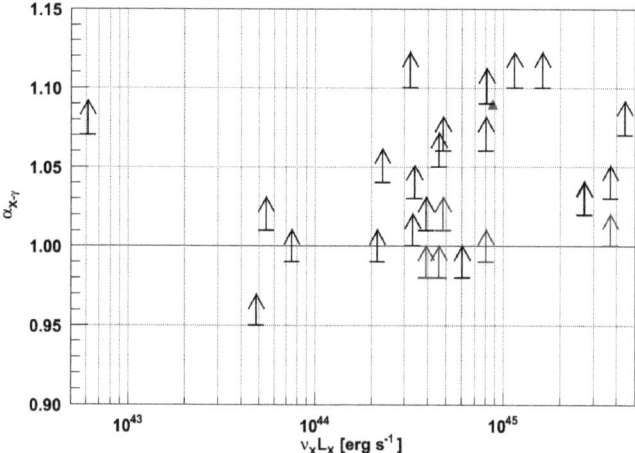

Figure 6.2: Broad-band spectral indices $\alpha_{X-\gamma}$ of the individual objects vs their X-ray luminosity $\nu_X L_X$. The VHE flux upper limits convert into lower limits for $\alpha_{X-\gamma}$. The majority of the objects have values larger than unity implying a higher energy output at 1 keV than at 200 GeV. The blue arrows indicate the lower limits found by Albert et al. (2008c). The red triangle marks the measured value for the stacked blazar sample.

lower limits and for the stacked sample measurement are listed in table 6.3. Assuming relative global errors of 5 % and 50 % for the X-ray and γ-ray fluxes, respectively, the errors $\delta_{X-\gamma}$ in $\alpha_{X-\gamma}$ become $\delta_{X-\gamma} = ^{+0.04}_{-0.02}$.

Spectral energy distribution

The spectral energy distribution is composed of the values listed in table 6.1 and the VHE γ-ray spectrum. In figure 6.3 the spectral luminosity νL_ν is plotted vs the frequency. The scattering in the radio, optical and X-ray bands is quite large, up to three orders of magnitude. This is due to the fact, that the objects have been chosen based mainly on the X-ray flux. The redshift range leads to a spread when transforming the SED into the jetframe.

Except for three objects the spectral indices at 1 keV indicate a declining spectrum. The mean spectral index at 1 keV is $\alpha_X = 2.44$. Therefore the mean synchrotron peak energy has to be less than 1 keV. Together with the mean optical luminosity at 640 nm it can be restricted roughly to values $E_{\text{sync}} \sim 10 - 100$ eV.

The estimation of the position of the VHE peak is more difficult. The measured spectrum

6 Steady state emission of blazars

Object	$\alpha_{X-\gamma}$
1ES 0033+595	1.04
1ES 0120+340	1.02
1ES 0229+200	1.00
RX J0319.8+1845	1.01
1ES 0323+022	1.05
1ES 0414+009	1.03
1RXS J044127.8+150455	1.10
1ES 0647+250	1.10
1ES 0806+524	1.06
1ES 0927+500	1.06
1ES 1028+511	1.07
RGB J1117+202	1.09
RXS J1136.5+6737	1.03
B2 1215+30	0.98
2E 1415.6+2557	1.10
PKS 1424+240	0.99
RX J1725.0+1152	1.07
1ES 1727+502	1.01
1ES 1741+196	0.99
B3 2247+381	0.95
Overall	1.09

Table 6.3: Lower limit values for the broad-band spectral indices $\alpha_{X-\gamma}$ of the BL Lac object sample. The majority of the lower limits exceeds unity indicating the trend towards a larger energy output in X-rays than in VHE γ-rays. The overall value is the measurement of the stacked sample.

at $z = 0$ indicates an inverse Compton peak at energies less than 100 GeV. However, the deabsorbed spectrum is still increasing beyond 1 TeV. As can be seen already in figure 6.1, the third bin of the spectrum has a large error on the flux value. Thus it cannot be excluded that the sample is somewhat twofold with a contribution of all objects to the lower energy side of the spectrum and a contribution only of the low-redshift objects to the higher energies. The absorption within the EBL would damp the high energy part of the high-redshift objects. In this case the high energy bins in the deabsorbed spectrum would be overestimated and the VHE peak shifted to lower energies. Recent measurements of BL Lac objects by FGST below ~ 100 GeV favour the latter scenario. They revealed mostly increasing spectra with flux levels higher than in the VHE regime indicating peak positions below ~ 100 GeV (Abdo et al., 2009).Due to the scattering within the optical and X-ray luminosities one can also expect a scattering of the individual synchrotron and inverse Compton peak energies of the order of 10–100. Regarding this large scattering, the determination of the peak positions of the stacked

6.1 Spectral characteristics

blazar sample is a good estimation.

In steady-state emission the energy densities of magnetic field and radiation are comparable. An inverse Compton peak position lying below $\sim 100\,\mathrm{GeV}$ is probably hiding the factor of 5.59 missing luminosity in the VHE regime needed for equality of the peaks.

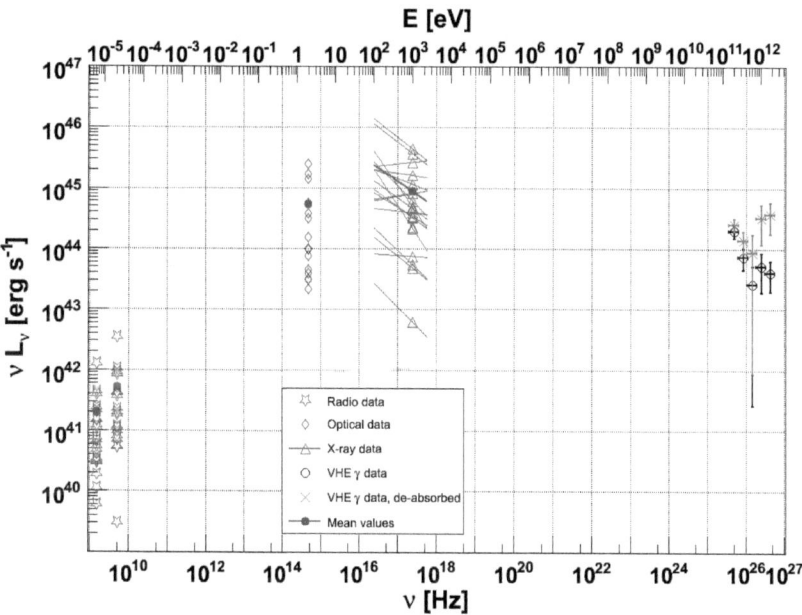

Figure 6.3: Spectral energy distribution of the stacked BL Lac object sample in the jetframe. Shown are the measurements in the radio (1.5 GHz and 5 GHz; stars), the optical (640 nm; diamonds) and the X-ray (1 keV; triangles) regime together with the X-ray spectral indices. The stacked VHE spectrum is shown as open circles and the deabsorbed one as crosses. The SED indicates a mean synchrotron peak at energies $\sim 10-100\,\mathrm{eV}$, while the inverse Compton peak cannot be determined unambiguously. The hard VHE spectrum favors a peak position above 1 TeV, but in the potential case of an overestimation of the high energy part of the deabsorbed spectrum allows for a peak energy well below 100 GeV.

6.2 Comparison with known steady state sources

In the last years several HBLs detected in the VHE regime have been claimed to be measured in a low emission state. In the following a selection of five of these sources will be presented and their spectral characteristics discussed. Finally the spectral energy distributions will be compared to the one of the stacked blazar sample.

6.2.1 HBLs measured in a low emission state

The majority of the sources listed in the following have redshifts $z < 0.05$ (exception: 1ES 1218+304, $z = 0.182$). Thus even low VHE emission states can easily be measured with current Cherenkov telescopes. These sources and their spectral modeling will serve as blueprints for the VHE emission of the stacked BL Lac object sample because all fulfill the selection criteria as described in section 5.1.3.

Mkn 421

As already denoted in chapter 5, Mkn 421 is the first source detected in the VHE range. In 2004 and 2005 it was observed by MAGIC in a low emission state (Albert et al., 2007c). The integral flux amounted to 0.5-2 times the Crab Nebula flux above 200 GeV. The SED with multi-wavelength data from KVA (optical), ASM/RXTE[2] (X-ray) and MAGIC could be well fitted with a leptonic one-zone synchrotron self-Compton model.

1ES 1218+304

1ES 1218+304 was discovered by MAGIC in 2005 (cf. chapter 5). Meyer (2008) reports about the analysis of observations performed in 2006 where the mean integral flux level above 150 GeV was lower by 50 % than in 2005. The result was a detection on the level of 4.0 standard deviations. In this case the question arises whether this source was really detected in a low (steady) emission state because the measured flux level was at the detection limit of MAGIC. Furthermore there was a hint on daily flux variability within the sample. However, the SED model fit will be included in this consideration for comparison.

Mkn 501

Mkn 501 was observed by MAGIC in 2006 within a multi-wavelength campaign together with KVA (optical) and the Suzaku satellite (X-rays). The source was detected in a low VHE emission state with an integral flux above 200 GeV of 0.2 times the Crab Nebula

---
[2]All Sky Monitor of the Rossi X-ray Timing Explorer

flux. As for Mkn 421, the SED could be fitted with a one-zone synchrotron self-Compton model which will be used in this work (Anderhub et al., 2009).

1ES 1959+650

During a multi-wavelength campaign in 2006, this source revealed one of the lowest VHE flux levels measured so far. The campaign included coverage of the optical (KVA, Tuorla optical telescope, Perugia AIT[3] and the Swift satellite), UV (Swift), soft X-ray (Swift and Suzaku) and hard X-ray (Suzaku) bands (Tagliaferri et al., 2008). Again the SED was fitted with a one-zone synchrotron self-Compton model. 1ES 1959+650 was discovered in 2002 by the Whipple Collaboration in a variable emission state with flux measurements up to 5 times the Crab Nebula flux (Holder et al., 2003).

1ES 2344+514

Observed with MAGIC in 2005 and 2006, 1ES 2344+514 was measured in a low VHE emission state (Albert et al., 2007b). The flux level was comparable to previous measurements by other Cherenkov telescopes (Badran et al., 2001; Aharonian et al., 2004).

Thus all sources listed above could be fitted with a leptonic SSC model which is the favoured model for these objects up to now.

6.2.2 Broad-band spectral indices

In table 6.4 the broad-band spectral indices $\alpha_{X-\gamma}$ for the five sources detected in a low flux state are listed. The X-ray flux at 1 keV has been determined from the model fit and the VHE one at 200 GeV from the deabsorbed energy spectrum. As for the undetected BL Lac object sample, 4 out of 5 objects have values $\alpha_{X-\gamma} \geq 1$. This strongly supports the claim of less luminosity in the VHE regime than in the X-ray band. In figure 6.4 the detected low state blazars are added to the lower limits from the BL Lac object sample. The result found here underlines the implications of the spectral blazar sequence with HBLs characterised by a higher energy output in the synchrotron component then at very high energies. Furthermore HBLs – compared to FSRQs – represent a late evolutionary scenario for blazars with a weak broad-band emission caused by the low particle densities in the jet and a VHE component at higher frequencies caused by the low optical depths near the black hole.

---

[3]Perugia Automatic Imaging Telescope

6 Steady state emission of blazars

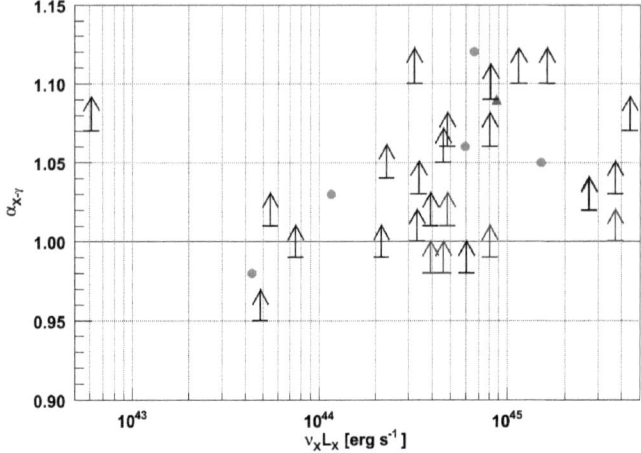

Figure 6.4: Same as figure 6.2 with the known low-state sources added as green points.

Source	$\alpha_{X-\gamma}$
Mkn 421	1.06
1ES 1218+304	1.05
Mkn 501	1.03
1ES 1959+650	1.12
1ES 2344+514	0.98

Table 6.4: Broad-band spectral indices $\alpha_{X-\gamma}$ for the low-state blazars.

6.2.3 Spectral energy distribution

For the comparison of the SED of the stacked BL Lac object sample with the ones of the sources detected in a low emission state, the model fits for the latter ones are transformed into luminosities. The comparison with the detected sources in a low emission state confirms the trend found for the stacked blazar sample. As can be seen in figure 6.5, the VHE component of all five sources match the stacked blazar spectrum within one order of magnitude at 200 GeV while the low energy component shows a large scatter at 1 keV, similar to the individual values of the undetected objects. In principle, the undetected objects do not differ from the detected ones. So it seems only logical to state that the lack of an individual detection is simply based on the low flux of these objects and only a stacked analysis made them detectable. The SED of 1ES 1218+340 might not be an ideal candidate for this comparison. As already discussed above, the SED of 1ES 1218+304 might not have been measured in the steady emission state,

6.2 Comparison with known steady state sources

which is supported by the higher luminosity level in the VHE range.

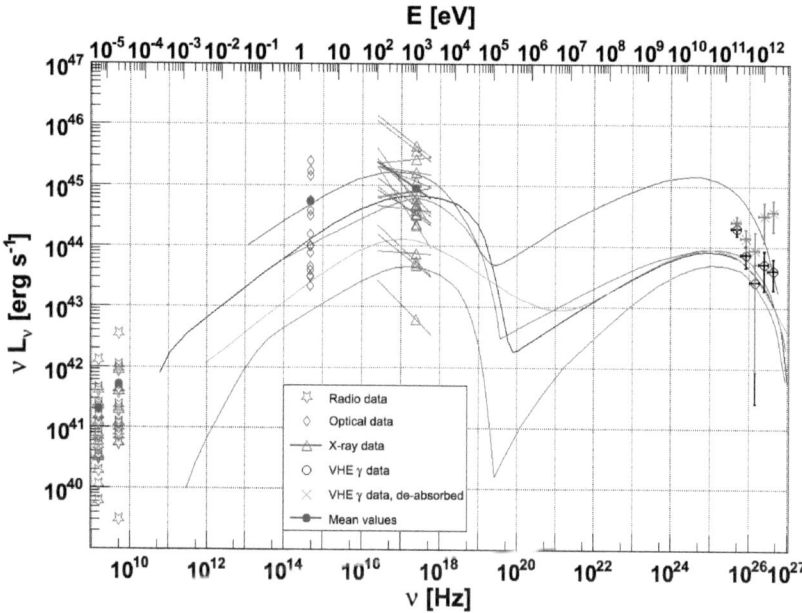

Figure 6.5: Spectral energy distribution of the stacked BL Lac object sample together with the single-zone SSC model fits of the detected low VHE emission state sources. Red: Mkn 421; blue: 1ES 1218+304; green: Mkn 501; black: 1ES 1959+650; pink: 1ES 2344+514.

6 Steady state emission of blazars

7 Conclusions and outlook

High-frequency peaked blazars represent the dominant population of extragalactic sources at very high energies. Improving our understanding of these enigmatic sources has been a prime motivation for building the MAGIC telescope, optimised for observing them at a low threshold energy, and hereby reducing the effect of γ-ray attenuation due to pair production in photon-photon interactions with the low-energy photons of the EBL. A large set of data has been piled up from the MAGIC blazar key science program, analysed here for the first time collectively.

So far, most of the blazars have been observed in a flaring state, resulting in a clear bias among the known VHE emitters. This observational bias has had a strong impact on the correlations found between fundamental scale parameters which determine the blazar properties, such as the so-called blazar sequence (Fossati et al., 1998). For 20 X-ray selected blazars which were not detected individually, upper limits on the integral flux on a 99.7% (3σ) confidence level have been calculated. They lie between 2.9% and 14.7% of the integral flux of the Crab Nebula in the VHE range.

The combination of the individual signals by means of a stacking method has led to the detection of a cumulative signal with a significance of 4.5 standard deviations in 349.5 hours effective exposure time (corresponding to 5.5 TBytes of raw data processed through the Würzburg data centre). The measured cumulative integral flux amounts to 1.4% of the Crab Nebula flux above 150 GeV and the differential spectral index is $\alpha = -3.15$, applying a simple power law fit to the data. None of the objects has shown variability on the time scales probed with the observations, i.e. the cumulative excess grows smoothly with t_{eff}. Therefore, the measured signal can be interpreted as the baseline emission of the sample.

Compared to the mean X-ray energy flux of the objects at 1 keV, the VHE flux at 200 GeV is significantly lower by a factor 5.59 implying a broad-band spectral index between the X-ray and VHE γ-ray regime $\alpha_{X-\gamma} = 1.09$. The lower limits of the broad-band spectral indices, converted from the VHE flux upper limits, show a clear tendency to be greater than unity.

7 Conclusions and outlook

Since the spectrum is steep, a bolometric comparison of the high-energy emission component with the lower energy synchrotron emission component is difficult to achieve in a model-independent way. Another source of uncertainty is the pair-production attenuation of the γ-rays. Within the uncertainties, the steady-state spectrum is in accordance with the synchrotron-self-Compton model for the emission from relativistic jets, and with the fact that the energy density of the synchrotron radiation field does not exceed the energy density of the magnetic field in steady state. This is expected for jets which are collimated by the magnetic field (Marscher et al., 2008).

To further support the idea that the stacked source signal indicates steady-state emission, the broad-band spectral indices and spectral energy distributions have been compared with the ones of five well-known blazars: Mkn 421, 1ES 1218+304, Mkn 501, 1ES 1959+650 and 1ES 2344+514. The result is that this control sample shows fair agreement with the "stacked source". So the objects within the sample seem to lie just below the sensitivity limit of the MAGIC telescope.

If the interpretation of the MAGIC stacked-source signal is correct, next generation Cherenkov telescope arrays such as CTA1[1] (Wagner et al., 2009b) will observe the steady-state emission of each object from the sample individually. Monitoring observations will allow to determine the duty cycle of the flares which have so far biased our knowledge of the spectra emitted by blazars. One of the sample objects, PKS 1424+240, has meanwhile turned into a flaring state detected by Cherenkov telescopes (Acciari et al. et al., 2009; Teshima, 2009).

The stacked spectrum can also help to determine the fraction of the diffuse extragalactic background produced by blazars more accurately. Inoue et al. (2010) have recently modeled the extragalactic γ-ray background assuming templates for the blazar spectral energy distribution obtained from the blazar sequence (Fossati et al., 1998) which is known to be affected by selection biases (Padovani et al., 2003), such as the one treated in this thesis. It is important to find out the remaining flux which can then be attributed to other putative sources, such as dark matter annihilation in the Universe.

---

[1] Cherenkov Telescope Array

A Data compendium

All data sequences which were used for the analysis can be found in the following sections. Listed will be the sequence number as found in the Würzburg data centre, the start time of the recorded data and the duration of the sequence. In addition the zenith distance, the mean pedestal RMS of the inner camera, the data rate after image cleaning and the inhomogeneity parameter are listed. Some sequences have been modified by removing single data runs due to quality reasons as described in section 5.3. These sequences are then marked with a *.

Sequence	Time	Duration [min]	Zd [°]	Ped. RMS	Data rate [Hz]	Inhom.
70006	2005-10-04 04:13:42	16.72	20–23	1.12	218	5.6
70015	2005-10-04 04:32:38	23.08	15–19	1.12	224	5.2
70025	2005-10-04 04:57:39	24.40	10–14	1.12	224	4.8
70035	2005-10-04 05:24:26	18.70	7–9	1.13	221	4.8
80433	2006-01-20 22:01:22	30.82	7–12	1.05	218	5.5
80447	2006-01-20 22:34:07	27.20	7–8	1.06	201	5.4
80752	2006-01-23 21:52:04	20.30	8–12	1.10	212	5.7
80764	2006-01-23 22:17:40	13.12	6–8	1.06	219	6.1
83987	2006-02-26 22:04:03	19.95	25–28	1.13	198	6.2
83996	2006-02-26 22:26:09	19.95	30–34	1.13	182	6.2
84227	2006-03-02 22:27:57	14.95	34–37	1.17	196	6.2

Table A.1: Data sequences of the Crab Nebula for the 2005/2006 dataset (300 MHz FADC system).

A Data compendium

Sequence	Time	Duration [min]	Zd [°]	Ped. RMS	Data rate [Hz]	Inhom.
101041	2006-09-21 03:45:45	38.90	32–40	1.27	137	15.1
101422	2006-09-25 03:17:25	58.02	31–43	1.31	163	6.8
103609	2006-10-21 04:14:33	97.42	7–16	1.22	191	8.4
105356	2006-11-19 02:52:03	18.58	7–8	1.21	189	7.3
109567	2006-12-25 00:10:44	58.10	7–10	1.19	176	8.5
109604	2006-12-25 02:14:24	42.32	24–33	1.21	159	10.0
109765	2006-12-26 02:11:08	43.67	25–33	1.24	149	7.3
112070	2007-01-21 20:01:49	58.25	24–36	1.21	167	6.2
112256	2007-01-22 21:05:04	58.22	10–22	1.26	161	7.5
112273	2007-01-22 22:07:49	19.58	7–9	1.18	167	7.6

Table A.2: Data sequences of the Crab Nebula for the 2006/2007 dataset (300 MHz FADC system with optical splitters).

Sequence	Time	Duration [min]	Zd [°]	Ped. RMS	Data rate [Hz]	Inhom.
216820	2007-02-13 21:30:39	105.93	8–30	1.04	189	6.9
218546	2007-02-15 21:32:23	88.25	9–28	1.05	189	5.3
225811	2007-03-07 21:09:26	17.75	20–25	1.11	190	6.3
225934	2007-03-08 20:16:10	65.95	10–24	1.16	156	6.7
228497	2007-03-16 20:45:25	22.48	22–28	1.17	189	5.7
318228	2008-01-01 21:53:07	61.00	16–30	1.00	203	6.9
319398	2008-01-03 21:46:10	63.60	15–30	1.00	200	7.0

Table A.3: Data sequences of the Crab Nebula for the 2007/2008 dataset (2 GHz MUX FADC system).

1ES 0033+595

Sequence	Time	Duration [min]	Zd [°]	Ped. RMS	Data rate [Hz]	Inhom.
*98043	2006-08-19 01:48:54	20.18	36–38	1.16	173	6.9
98051	2006-08-19 02:11:43	13.38	35–35	1.16	168	9.0
*98553	2006-08-22 04:52:01	36.97	33–35	1.16	177	4.3
*98697	2006-08-23 04:46:58	38.47	33–36	1.15	174	4.4
98820	2006-08-24 03:46:26	7.50	31–31	1.15	179	8.5
98829	2006-08-24 04:04:30	25.03	32–32	1.14	176	7.9
*98840	2006-08-24 04:33:18	37.48	32–36	1.13	165	8.3
*99636	2006-09-01 03:15:06	60.60	31–33	1.13	168	4.1
*99754	2006-09-02 03:06:36	134.23	31–39	1.14	150	6.3
1001041	2008-07-09 04:25:44	20.00	36–38	1.19	137	10.6
1001171	2008-07-12 04:14:02	62.58	33–38	1.26	158	9.4

1ES 0120+340

Sequence	Time	Duration [min]	Zd [°]	Ped. RMS	Data rate [Hz]	Inhom.
63776	2005-08-04 04:55:52	10.25	9–12	0.93	231	7.0
64025	2005-08-05 04:41:57	24.20	9–12	0.92	226	6.6
64713	2005-08-08 04:33:03	30.98	6–13	0.91	217	7.3
64943	2005-08-09 04:21:42	45.13	6–13	0.93	225	6.6
65142	2005-08-10 04:23:15	46.45	6–13	0.90	227	7.3
65333	2005-08-11 04:05:54	38.45	9–15	0.90	226	7.2
65356	2005-08-11 04:50:12	12.00	6–7	0.90	229	6.2
65531	2005-08-12 03:59:36	51.15	6–16	0.89	230	6.8
65563	2005-08-12 05:05:09	9.43	6–6	0.90	232	6.2
65695	2005-08-13 03:50:21	44.83	8–16	0.93	216	7.8
65716	2005-08-13 04:43:09	30.12	6–8	0.92	211	7.2
65778	2005-08-14 04:28:54	44.70	6–10	0.91	224	6.6
65860	2005-08-15 03:37:48	59.23	7–18	0.92	224	7.5
65898	2005-08-15 04:44:49	29.90	6–7	0.93	225	6.6
65929	2005-08-16 03:40:27	59.88	6–17	0.95	224	7.8
65959	2005-08-16 04:49:34	26.43	6–7	0.97	227	7.4
65976	2005-08-17 03:51:48	81.75	6–13	0.92	222	6.7

Table A.4: Data sequences of 1ES 0033+595 and 1ES 0120+340.

A Data compendium

1ES 0229+200

Sequence	Time	Duration [min]	Zd [°]	Ped. RMS	Data rate [Hz]	Inhom.
98394	2006-08-21 03:35:40	47.53	21–31	1.08	197	5.0
101749	2006-09-27 03:26:22	55.53	8–16	1.08	193	6.3
101768	2006-09-27 04:26:54	66.20	18–32	1.10	183	6.6
101911	2006-09-28 03:24:42	57.93	9–18	1.10	175	7.2
*101929	2006-09-28 04:26:27	77.42	18–35	1.14	165	6.6
102290	2006-10-01 03:23:59	58.17	10–19	1.18	165	5.4
*103142	2006-10-20 01:48:41	117.57	8–30	1.06	203	7.6
103375	2006-10-21 01:51:39	37.37	9–13	1.11	185	6.8
*103584	2006-10-21 03:09:54	57.98	20–33	1.10	175	7.3

RX J0319.8+1845

Sequence	Time	Duration [min]	Zd [°]	Ped. RMS	Data rate [Hz]	Inhom.
69286	2005-09-29 02:38:17	20.40	18–21	1.05	212	8.0
69301	2005-09-29 03:09:38	6.52	14–14	1.06	213	8.1
69807	2005-10-03 02:18:21	22.07	17–21	1.01	162	9.7
69815	2005-10-03 02:42:26	25.97	13–17	1.01	160	8.2
69824	2005-10-03 03:11:03	19.50	10–12	1.01	169	7.9
69831	2005-10-03 03:33:44	18.30	10–10	1.00	172	7.4
69838	2005-10-03 03:54:21	17.63	11–12	0.99	170	7.3
70133	2005-10-05 02:09:08	20.77	18–22	1.06	218	6.9
70142	2005-10-05 02:31:56	23.85	13–17	1.06	220	6.6
70153	2005-10-05 02:57:47	20.17	11–13	1.05	218	7.8
70162	2005-10-05 03:19:57	18.58	10–10	1.02	221	7.9
70170	2005-10-05 03:40:50	22.75	10–12	1.03	218	6.7
80727	2006-01-23 20:36:58	20.20	11–13	0.99	242	4.2
80738	2006-01-23 20:59:29	28.02	14–19	1.03	225	4.0

Table A.5: Data sequences of 1ES 0229+200 and RX J0319.8+1845.

1ES 0323+022

Sequence	Time	Duration [min]	Zd [°]	Ped. RMS	Data rate [Hz]	Inhom.
69474	2005-09-30 02:30:57	19.23	31–33	1.10	181	8.1
69483	2005-09-30 02:54:44	19.55	29–31	1.12	186	8.0
69491	2005-09-30 03:16:32	23.45	27–28	1.11	190	7.3
69502	2005-09-30 03:43:27	16.02	26–27	1.10	191	7.4
69510	2005-09-30 04:01:59	18.03	26–27	1.10	191	7.1
69518	2005-09-30 04:21:41	2.67	27–27	1.10	189	7.1
69959	2005-10-04 02:14:53	15.93	32–34	1.00	215	6.9
69967	2005-10-04 02:33:11	22.23	29–31	0.99	221	7.2
69976	2005-10-04 02:57:18	24.57	27–29	0.98	224	6.4
69987	2005-10-04 03:23:44	21.12	26–27	0.98	223	6.2
69997	2005-10-04 03:46:53	20.83	26–27	0.97	221	6.6
70731	2005-10-13 02:44:54	18.12	26–27	1.01	213	7.3
70739	2005-10-13 03:08:48	22.63	26–27	1.00	209	7.2
71105	2005-10-26 00:30:55	14.25	34–36	0.99	199	6.5
71112	2005-10-26 00:47:31	17.80	32–34	1.00	200	6.6
71165	2005-10-28 01:24:35	16.65	27–28	0.98	230	7.6
71173	2005-10-28 01:43:26	21.13	26–27	0.99	227	7.0
71352	2005-10-29 00:30:26	31.65	30–34	0.99	210	8.4
71366	2005-10-29 01:03:59	24.07	28–30	0.99	215	8.2
71377	2005-10-29 01:30:01	22.83	26–28	1.00	216	7.7
71510	2005-10-31 01:21:50	28.35	26–28	1.07	190	7.3
71679	2005-11-01 00:18:47	21.08	32–34	0.94	204	8.3
71688	2005-11-01 00:41:52	20.70	29–32	0.96	209	7.9
71697	2005-11-01 01:04:35	19.63	27–29	0.97	214	7.9
71706	2005-11-01 01:26:05	14.08	27–27	0.99	211	7.6
71936	2005-11-02 00:11:05	22.55	32–36	0.93	204	8.1
71946	2005-11-02 00:35:41	22.65	29–31	0.92	207	7.5
71962	2005-11-02 01:00:44	20.37	28–29	0.93	209	8.2
71971	2005-11-02 01:23:13	15.92	26–27	0.94	209	8.1
75753	2005-11-24 23:03:53	26.62	29–32	1.02	182	8.8
75768	2005-11-24 23:32:29	36.50	26–29	1.05	186	8.5
76128	2005-11-30 22:59:14	25.77	27–29	0.96	200	8.4
76139	2005-11-30 23:27:28	19.95	26–27	0.96	198	9.1

Table A.6: Data sequences of 1ES 0323+022.

A Data compendium

1ES 0414+009

Sequence	Time	Duration [min]	Zd [°]	Ped. RMS	Data rate [Hz]	Inhom.
72012	2005-11-02 01:43:32	19.57	29–31	0.95	208	8.3
72021	2005-11-02 02:05:19	22.13	28–29	0.95	207	7.6
72036	2005-11-02 02:29:44	22.25	28–28	1.01	206	7.1
72047	2005-11-02 02:54:17	20.17	28–29	0.95	205	6.7
72056	2005-11-02 03:16:26	23.25	29–31	0.97	202	6.8
72066	2005-11-02 03:41:32	23.18	31–34	0.98	199	6.9
74319	2005-11-05 02:59:16	23.93	29–30	1.04	210	7.7
74881	2005-11-08 02:36:53	23.15	28–29	0.95	188	9.4
74891	2005-11-08 03:02:05	23.02	30–32	0.95	186	9.2
74901	2005-11-08 03:27:53	15.17	32–34	0.94	182	8.6
75271	2005-11-11 02:48:26	20.50	30–32	1.01	213	7.8
75282	2005-11-11 03:14:28	16.53	32–34	1.00	208	7.4
75793	2005-11-25 00:12:20	25.57	29–31	1.05	181	8.9
75803	2005-11-25 00:40:23	23.73	28–29	1.05	184	8.4
75813	2005-11-25 01:06:23	20.50	28–28	1.02	186	8.3
76148	2005-11-30 23:52:55	22.48	29–30	0.97	188	8.5
76157	2005-12-01 00:20:04	20.05	28–28	0.97	188	7.8
76165	2005-12-01 00:42:44	27.40	28–28	0.97	192	7.9
76176	2005-12-01 01:12:35	17.33	28–29	0.96	181	7.9
76432	2005-12-03 00:43:25	21.58	28–28	1.02	181	8.1
76441	2005-12-03 01:07:02	26.95	28–30	1.04	180	7.9
76869	2005-12-05 00:36:41	23.35	28–28	0.93	193	8.2
76879	2005-12-05 01:02:05	22.92	29–30	0.97	187	7.7
77272	2005-12-07 00:36:34	18.42	28–29	0.96	192	8.4
77281	2005-12-07 00:57:08	20.82	29–30	0.97	192	8.1
78105	2005-12-22 21:45:55	21.48	33–36	1.03	173	8.6
78117	2005-12-22 22:09:29	22.00	30–33	1.02	175	8.3
78130	2005-12-22 22:34:49	22.38	28–30	1.04	179	7.7
78145	2005-12-22 22:59:13	21.07	28–28	1.03	182	7.5
78160	2005-12-22 23:22:23	18.10	28–28	1.03	181	7.1
78386	2005-12-24 21:31:37	20.22	34–36	1.02	179	7.9
78394	2005-12-24 21:53:46	20.47	31–34	1.02	183	7.0
78403	2005-12-24 22:17:10	17.90	29–30	1.03	182	7.2
78411	2005-12-24 22:37:04	18.03	28–29	1.03	181	6.8
78419	2005-12-24 22:57:28	17.52	28–28	1.02	184	6.6

Table A.7: Data sequences of 1ES 0414+009.

1ES 0414+009 continued

Sequence	Time	Duration [min]	Zd [°]	Ped. RMS	Data rate [Hz]	Inhom.
78428	2005-12-24 23:16:58	14.10	28–28	1.03	183	6.3
78504	2005-12-26 21:28:49	19.35	34–36	1.03	182	8.1
78513	2005-12-26 21:50:22	18.92	31–32	1.02	187	8.2
78521	2005-12-26 22:11:40	25.10	31–31	1.03	182	7.9
78532	2005-12-26 22:39:08	23.78	28–28	1.01	183	6.7
78542	2005-12-26 23:05:01	18.98	28–28	1.00	185	7.1
79014	2005-12-30 21:24:37	20.57	32–35	1.05	163	9.2
79035	2005-12-30 21:47:35	16.47	29–31	1.07	167	8.5
79054	2005-12-30 22:18:43	15.35	28–29	1.01	165	8.7
79061	2005-12-30 22:36:16	25.93	28–28	0.99	170	9.4
79548	2006-01-01 21:06:58	19.18	33–35	0.96	169	9.8
79556	2006-01-01 21:28:11	19.50	31–33	0.96	173	9.0
79565	2006-01-01 21:51:01	25.28	28–30	0.95	174	9.0
79576	2006-01-01 22:42:10	17.32	28–28	0.93	178	8.9
80322	2006-01-19 20:10:19	18.65	31–33	1.01	178	8.5
80331	2006-01-19 20:30:58	19.23	30–31	1.00	181	7.8
80341	2006-01-19 20:52:28	21.92	28–29	1.00	183	7.8
80350	2006-01-19 21:17:01	19.15	28–28	1.00	173	6.9

1RXS J044127.8+150455

Sequence	Time	Duration [min]	Zd [°]	Ped. RMS	Data rate [Hz]	Inhom.
287148	2007-10-08 03:44:18	53.37	14–20	1.01	218	4.5
288100	2007-10-10 03:49:03	38.97	14–18	0.97	182	7.0
288155	2007-10-10 05:30:21	37.30	18–24	1.05	176	6.2
289127	2007-10-14 04:42:18	85.25	14–28	1.06	194	6.8
290121	2007-10-16 03:07:19	153.47	14–30	0.98	205	6.2
290601	2007-10-17 03:06:03	11.78	19–21	1.01	218	12.9
291340	2007-10-19 02:56:20	32.07	16–21	0.99	220	6.6
293591	2007-10-24 05:14:32	56.32	25–36	1.07	191	7.2
295270	2007-11-05 00:43:50	103.85	15–34	1.03	211	7.3
295820	2007-11-06 01:09:50	56.17	17–28	1.00	216	7.3
296317	2007-11-07 00:33:05	89.10	17–34	1.08	128	7.4
299489	2007-11-13 00:05:50	89.23	17–35	1.04	204	4.7

Table A.8: Data sequences of 1ES 0414+009 and 1RXS J044127.8+150455.

A Data compendium

1RXS J044127.8+150455 continued

Sequence	Time	Duration [min]	Zd [°]	Ped. RMS	Data rate [Hz]	Inhom.
300256	2007-11-14 00:01:05	88.80	17–35	1.00	280	4.9
301051	2007-11-14 23:55:51	90.47	17–35	1.02	273	4.9
303505	2007-12-01 23:19:35	116.23	14–29	0.92	222	6.5
304148	2007-12-02 23:12:35	168.50	13–30	1.00	313	3.7
304929	2007-12-03 23:12:36	219.50	13–32	0.99	253	4.9
305519	2007-12-04 23:32:05	17.88	21–24	0.96	222	7.2
306561	2007-12-05 23:13:35	78.00	14–27	0.99	318	4.0

1ES 0647+250

Sequence	Time	Duration [min]	Zd [°]	Ped. RMS	Data rate [Hz]	Inhom.
333734	2008-02-04 20:44:53	207.50	3–31	0.99	218	7.1
334384	2008-02-05 20:38:52	212.22	3–32	1.00	214	7.3
335038	2008-02-06 20:37:08	73.38	9–31	1.02	180	8.4
335277	2008-02-07 20:32:23	197.98	4–31	1.03	207	7.0
336718	2008-02-27 21:34:26	24.95	3–7	1.06	214	5.4
336746	2008-02-27 22:02:54	12.17	8–10	1.06	211	5.8
336761	2008-02-27 22:20:10	57.47	11–24	1.07	206	4.4
337132	2008-02-28 20:26:24	169.63	4–25	1.08	165	5.8
338517	2008-03-03 21:08:41	63.22	4–14	1.10	171	6.5
338752	2008-03-04 20:31:09	166.45	3–30	1.09	141	7.0
339185	2008-03-05 21:41:25	95.72	9–30	1.07	147	6.0
339520	2008-03-06 21:43:10	89.77	10–30	1.10	190	4.9
340008	2008-03-07 20:31:55	154.30	3–30	1.12	191	5.7
340545	2008-03-08 20:32:10	59.25	4–10	1.10	204	6.5
340606	2008-03-08 21:43:11	84.18	11–30	1.10	196	5.7
341125	2008-03-09 21:26:24	85.37	9–28	1.07	199	5.6

Table A.9: Data sequences of 1RXS J044127.8+150455 and 1ES 0647+250.

1ES 0806+524

Sequence	Time	Duration [min]	Zd [°]	Ped. RMS	Data rate [Hz]	Inhom.
71567	2005-10-31 04:20:56	23.12	32–34	1.07	181	7.3
71576	2005-10-31 04:45:59	25.30	29–32	1.05	186	6.9
71586	2005-10-31 05:14:53	9.85	28–29	1.05	191	7.2
72079	2005-11-02 04:18:34	16.90	32–34	0.96	192	7.7
72087	2005-11-02 04:38:08	21.32	30–32	0.96	200	7.9
74524	2005-11-06 03:58:44	21.18	33–35	0.98	191	8.3
74533	2005-11-06 04:21:52	22.83	29–31	0.99	190	7.4
74543	2005-11-06 04:46:56	23.68	27–29	0.97	198	7.1
74553	2005-11-06 05:12:40	4.82	26–26	0.98	196	6.2
74557	2005-11-06 05:20:49	21.40	24–26	0.98	199	6.3
74567	2005-11-06 05:44:02	32.02	24–24	0.98	198	6.6
74911	2005-11-08 03:55:16	24.28	31–34	0.90	178	7.3
74921	2005-11-08 04:21:46	23.10	29–31	0.87	182	8.0
74930	2005-11-08 04:47:41	22.33	26–28	0.88	182	7.4
74939	2005-11-08 05:12:17	23.40	25–26	0.87	187	7.8
74949	2005-11-08 05:37:38	22.38	24–24	0.88	186	7.7
74958	2005-11-08 06:02:10	17.73	24–24	0.92	193	7.5
75290	2005-11-11 03:35:17	24.23	33–36	0.96	195	7.9
76188	2005-12-01 05:20:35	24.53	25–27	0.96	173	8.6
76197	2005-12-01 05:47:17	24.53	27–29	0.98	165	7.7
76474	2005-12-03 02:50:05	26.08	28–30	0.97	181	8.1
76484	2005-12-03 03:28:08	22.93	25–26	0.99	179	7.6
76494	2005-12-03 03:54:50	25.27	24–24	1.00	167	7.5
76505	2005-12-03 04:22:35	24.23	24–24	0.98	175	7.7
76523	2005-12-03 05:32:20	22.65	26–28	0.97	164	8.0
76943	2005-12-05 03:33:01	21.93	24–25	0.96	183	8.3
76959	2005-12-05 03:57:05	22.90	24–24	0.96	188	8.6
76969	2005-12-05 04:22:02	22.97	24–24	0.97	179	10.3
76979	2005-12-05 04:50:08	19.83	24–25	0.95	187	8.7
76991	2005-12-05 05:12:29	22.48	26–27	0.97	182	9.3
77001	2005-12-05 05:36:55	23.08	27–29	0.97	178	8.0
77010	2005-12-05 06:01:59	8.88	30–31	0.99	173	8.8
77015	2005-12-05 06:12:47	6.88	32–32	1.00	160	8.7
77346	2005-12-07 03:24:40	23.32	24–26	0.93	188	8.4
77356	2005-12-07 03:50:07	22.87	24–24	0.94	191	9.4

Table A.10: Data sequences of 1ES 0806+524.

A Data compendium

1ES 0806+524 continued

Sequence	Time	Duration [min]	Zd [°]	Ped. RMS	Data rate [Hz]	Inhom.
77367	2005-12-07 04:15:04	12.37	24–24	0.94	192	8.2
77394	2005-12-07 04:32:26	4.72	24–24	0.95	195	8.1
77443	2005-12-07 04:40:43	4.28	24–24	0.94	198	8.0
77447	2005-12-07 04:47:10	22.85	24–26	0.96	191	9.5
77457	2005-12-07 05:12:04	7.93	26–26	0.94	193	8.9
77634	2005-12-09 02:38:23	16.63	27–28	0.93	191	7.7
77642	2005-12-09 02:57:17	17.75	26–27	0.93	191	8.1
77651	2005-12-09 03:16:55	18.12	24–25	0.95	190	7.7
77660	2005-12-09 03:37:04	17.93	24–24	0.90	199	8.2
77670	2005-12-09 03:57:02	18.00	24–24	0.95	191	9.3
77678	2005-12-09 04:16:58	18.03	24–24	0.91	195	8.5
77686	2005-12-09 04:37:04	17.97	24–25	0.95	191	9.9
77694	2005-12-09 04:56:55	18.07	25–26	0.95	191	8.9
77702	2005-12-09 05:17:04	17.88	27–29	0.97	190	8.7
77710	2005-12-09 05:36:52	18.17	29–30	0.97	185	8.1
78583	2005-12-27 01:29:55	18.95	27–28	1.02	181	7.3
78591	2005-12-27 01:50:58	19.33	25–27	1.02	188	7.1
78599	2005-12-27 02:12:25	16.72	24–25	1.03	188	7.1
78606	2005-12-27 02:31:13	20.13	24–24	1.03	185	6.8

1ES 0927+500

Sequence	Time	Duration [min]	Zd [°]	Ped. RMS	Data rate [Hz]	Inhom.
78615	2005-12-27 02:56:22	19.02	24–26	1.01	187	7.4
78623	2005-12-27 03:17:43	15.10	23–24	1.01	186	7.5
78631	2005-12-27 03:39:41	35.07	21–22	1.01	189	7.0
78872	2005-12-30 02:46:40	21.42	24–26	1.02	182	8.1
78881	2005-12-30 03:10:02	19.25	22–24	1.02	181	8.2
78889	2005-12-30 03:31:32	18.95	21–22	1.03	180	7.8
78897	2005-12-30 03:53:01	21.10	21–21	1.02	180	8.2
78905	2005-12-30 04:16:25	18.73	21–22	1.01	179	7.5
79168	2005-12-31 02:41:10	12.60	25–26	1.02	165	8.7
79175	2005-12-31 02:55:55	19.80	23–24	1.03	168	8.1
79185	2005-12-31 03:19:28	20.62	22–23	1.03	171	8.3

Table A.11: Data sequences of 1ES 0806+524 and 1ES 0927+500.

1ES 0927+500 continued

Sequence	Time	Duration [min]	Zd [°]	Ped. RMS	Data rate [Hz]	Inhom.
79196	2005-12-31 03:43:13	22.47	21–21	1.04	172	8.3
79206	2005-12-31 04:07:49	20.02	21–21	1.04	174	8.2
79466	2006-01-01 03:52:34	18.22	21–21	0.96	170	7.2
80075	2006-01-06 02:21:31	19.98	24–26	0.91	188	7.5
80083	2006-01-06 02:43:16	18.72	22–23	0.92	190	7.7
80091	2006-01-06 03:04:04	21.50	21–22	0.92	189	7.3
80100	2006-01-06 03:27:40	18.27	21–21	0.93	191	7.5
80110	2006-01-06 03:47:49	21.68	21–22	0.92	183	7.4
80652	2006-01-23 01:38:05	9.02	23–23	0.98	172	3.3
80657	2006-01-23 01:49:29	10.58	22–23	0.97	172	3.6
80899	2006-01-29 01:01:04	19.02	23–24	0.89	226	7.3
80910	2006-01-29 01:22:31	34.90	21–23	0.88	221	7.5
81358	2006-01-31 01:23:00	20.40	21–22	0.90	220	6.2
81368	2006-01-31 01:45:43	20.37	21–21	0.90	216	6.2
81726	2006-02-02 01:05:22	13.82	22–23	0.95	218	5.7
81733	2006-02-02 01:21:36	13.72	21–22	0.95	216	5.5
81740	2006-02-02 01:38:10	28.45	21–21	0.95	214	5.5
81948	2006-02-03 01:03:49	20.57	22–23	0.97	225	6.0
81958	2006-02-03 01:26:39	24.93	21–21	0.98	219	5.8
82133	2006-02-04 00:49:55	27.85	22–23	0.95	224	5.4
82146	2006-02-04 01:20:37	27.92	21–22	0.93	220	6.3
82918	2006-02-19 23:51:24	19.92	22–23	0.91	223	6.2
82928	2006-02-20 00:13:52	39.80	21–22	0.95	216	6.6
83195	2006-02-21 23:43:06	19.93	22–23	0.91	230	6.2
83205	2006-02-22 00:05:34	26.83	21–22	0.90	230	6.7
83387	2006-02-22 23:37:36	14.93	22–23	0.93	223	7.1
83395	2006-02-22 23:55:03	29.87	21–22	0.93	220	7.5
83591	2006-02-23 23:35:45	14.95	22–23	0.91	215	5.6
83599	2006-02-23 23:52:37	44.80	21–22	0.91	209	6.2
83771	2006-02-24 23:33:03	24.90	22–23	0.93	221	6.3
83783	2006-02-25 00:00:12	24.92	21–22	0.93	216	6.1
84015	2006-02-26 23:28:40	19.92	22–23	0.92	220	6.4
84026	2006-02-26 23:50:54	19.92	21–21	0.92	214	7.0
84241	2006-03-02 23:08:13	24.93	22–23	0.99	227	6.2
84253	2006-03-02 23:35:43	24.93	21–22	0.99	218	6.0

Table A.12: Data sequences of 1ES 0927+500.

A Data compendium

1ES 1028+511

Sequence	Time	Duration [min]	Zd [°]	Ped. RMS	Data rate [Hz]	Inhom.
226096	2007-03-08 22:45:28	36.43	26–31	0.99	173	5.8
*226308	2007-03-09 22:43:13	63.97	24–31	0.97	174	6.5
*226581	2007-03-10 22:41:10	61.53	24–30	0.92	185	7.2
227192	2007-03-11 22:32:10	74.47	23–31	0.97	195	7.0
227882	2007-03-12 22:58:57	36.25	24–27	0.93	197	6.8
231735	2007-04-13 20:56:39	80.15	22–27	0.88	208	6.8
233180	2007-04-15 20:58:56	74.90	22–25	0.90	204	6.8
238476	2007-05-04 21:25:12	7.98	23–23	0.94	209	9.0
238491	2007-05-04 21:36:57	26.40	23–26	0.99	200	8.2
*238602	2007-05-05 21:07:58	84.87	22–30	0.95	194	7.1
*239090	2007-05-07 21:19:27	67.00	23–30	0.97	143	7.3
305395	2007-12-04 05:26:05	57.55	23–28	1.14	192	5.9
306176	2007-12-05 05:47:37	55.47	22–25	1.06	199	6.2
307741	2007-12-07 06:01:50	28.00	22–23	0.89	220	7.1
308934	2007-12-08 05:49:50	37.22	22–24	0.90	222	6.6
310142	2007-12-09 05:42:50	43.57	22–24	0.90	216	6.0
311065	2007-12-10 05:43:06	48.23	22–24	0.96	175	5.2
315723	2007-12-21 05:26:05	49.47	22–23	0.95	192	7.0
320807	2008-01-05 03:58:53	94.40	22–24	0.86	213	6.4
321736	2008-01-06 03:58:38	149.95	22–32	0.84	218	6.7
322580	2008-01-07 03:45:07	37.20	22–24	0.90	199	6.3
322702	2008-01-07 05:59:38	46.27	27–32	0.90	184	5.9
323374	2008-01-08 02:48:22	60.82	23–30	0.87	219	6.2
323560	2008-01-08 06:00:52	41.55	27–33	0.88	215	6.6
325190	2008-01-10 03:34:06	28.73	22–24	0.94	214	7.1
325248	2008-01-10 04:31:38	67.25	22–26	0.91	214	6.8
325347	2008-01-10 06:09:37	27.13	30–33	0.88	210	7.0
325937	2008-01-11 05:23:07	1.95	25–25	0.83	218	9.4
325992	2008-01-11 06:21:07	22.58	31–35	0.83	212	8.0
326635	2008-01-12 03:11:09	45.35	22–25	0.88	220	7.4
326745	2008-01-12 04:59:53	47.08	24–28	0.88	218	6.7
326829	2008-01-12 06:25:37	21.68	33–35	0.86	207	7.6
327476	2008-01-13 03:15:24	110.30	22–24	0.88	202	6.9
327642	2008-01-13 06:04:11	35.05	31–35	0.85	211	7.2
328110	2008-01-14 05:29:23	69.48	27–36	0.87	209	7.6

Table A.13: Data sequences of 1ES 1028+511.

1ES 1028+511 continued

Sequence	Time	Duration [min]	Zd [°]	Ped. RMS	Data rate [Hz]	Inhom.
328544	2008-01-15 02:08:08	217.00	22–31	0.87	215	7.1
336420	2008-02-26 23:27:40	50.90	25–30	0.97	207	5.7
336820	2008-02-27 23:26:25	108.13	22–30	0.95	197	4.9
337293	2008-02-29 01:14:25	57.60	22–25	0.95	174	5.3

RGB J1117+202

Sequence	Time	Duration [min]	Zd [°]	Ped. RMS	Data rate [Hz]	Inhom.
*111804	2007-01-20 02:15:35	58.35	19–31	1.05	175	9.7
111824	2007-01-20 03:17:20	38.53	12–19	1.05	180	10.3
111989	2007-01-21 01:50:34	58.37	24–36	1.14	167	9.0
112006	2007-01-21 02:52:34	58.08	12–22	1.13	176	10.7
112156	2007-01-22 01:57:34	58.38	21–33	1.06	166	9.4
112174	2007-01-22 02:59:34	46.58	12–21	1.07	176	10.6
*112333	2007-01-23 01:36:20	58.37	26–38	1.14	163	8.7
*112352	2007-01-23 02:39:05	63.10	12–23	1.11	170	11.0
*112495	2007-01-24 01:46:49	124.83	9–34	1.12	142	5.4
*112530	2007-01-24 04:10:53	137.92	9–35	1.05	160	5.7
342047	2008-03-11 23:40:25	57.50	11–22	1.09	187	6.0
342103	2008-03-12 00:42:40	58.52	8–12	1.09	186	3.3
342504	2008-03-13 00:57:39	38.07	8–12	1.06	163	4.8
*342538	2008-03-13 01:39:41	82.15	12–28	1.05	145	4.5
342803	2008-03-14 01:27:25	118.43	11–35	1.05	160	5.4

Table A.14: Data sequences of 1ES 1028+511 and RGB J1117+202.

A Data compendium

RXS J1136.5+6737

Sequence	Time	Duration [min]	Zd [°]	Ped. RMS	Data rate [Hz]	Inhom.
*214984	2007-02-11 00:47:38	96.82	40–46	0.89	165	7.4
*215582	2007-02-12 00:44:40	156.60	39–45	0.90	169	7.3
216259	2007-02-13 00:38:26	65.30	41–46	0.90	184	6.9
217053	2007-02-14 00:39:23	55.20	41–45	0.87	180	7.0
*217883	2007-02-15 00:31:55	62.87	41–46	0.90	186	6.8
218767	2007-02-16 00:25:54	63.97	41–46	0.90	170	7.3
*219583	2007-02-17 00:25:23	61.97	41–45	0.92	164	6.5
*220336	2007-02-18 00:20:23	60.47	41–45	0.92	184	6.4
221183	2007-02-19 00:17:23	60.48	41–45	0.89	196	7.3
*221768	2007-02-20 00:09:10	220.05	39–45	0.95	163	6.8
*222436	2007-02-21 00:08:08	150.27	39–45	0.95	201	6.9

B2 1215+30

Sequence	Time	Duration [min]	Zd [°]	Ped. RMS	Data rate [Hz]	Inhom.
226671	2007-03-11 00:01:15	74.27	12–29	0.94	196	7.6
227303	2007-03-11 23:55:41	130.77	1–30	0.96	203	7.5
*227940	2007-03-12 23:47:15	22.02	26–30	0.95	192	7.5
229039	2007-03-23 01:59:39	34.35	6–15	1.11	218	4.4
230654	2007-04-11 23:12:25	38.07	4–13	0.85	215	8.0
230760	2007-04-12 00:38:25	55.92	7–20	0.86	216	6.5
*234958	2007-04-19 00:13:12	75.42	7–24	0.94	143	5.9
235550	2007-04-20 00:08:12	75.05	7–24	0.86	206	6.0
*236242	2007-04-21 00:08:27	70.95	8–23	0.93	199	6.5
332765	2008-02-02 01:53:52	62.28	22–37	0.87	204	8.2
333378	2008-02-04 01:40:22	66.57	23–37	0.89	204	8.8
339607	2008-03-06 23:22:25	171.02	2–40	0.97	194	6.1
340189	2008-03-07 23:57:40	136.17	2–31	0.97	186	5.8

Table A.15: Data sequences of RXS J1136.5+6737 and B2 1215+30.

2E 1415.6+2557

Sequence	Time	Duration [min]	Zd [°]	Ped. RMS	Data rate [Hz]	Inhom.
225035	2007-02-26 04:00:23	130.33	3–17	0.87	212	6.5
225466	2007-02-27 05:31:39	45.42	7–17	0.89	210	6.8
*228287	2007-03-16 03:21:25	126.32	3–23	0.96	164	8.1
*231569	2007-04-13 03:20:54	45.65	17–28	0.93	171	5.4
232147	2007-04-14 01:35:24	165.45	3–33	0.92	203	6.8
*232953	2007-04-15 02:28:39	34.97	8–15	0.88	212	6.4
*233006	2007-04-15 03:10:24	78.55	16–34	0.88	205	7.0
*233654	2007-04-16 02:24:25	56.05	8–20	0.88	205	7.0
233744	2007-04-16 03:40:40	56.08	24–36	0.94	194	6.5
233890	2007-04-17 02:37:10	83.52	11–35	0.88	214	5.7
234425	2007-04-18 01:36:42	74.98	3–16	0.84	212	6.7
*234549	2007-04-18 02:59:11	78.47	16–34	0.85	201	7.1
*240259	2007-05-11 01:19:12	63.13	14–29	0.98	213	6.5
240369	2007-05-11 02:29:56	5.37	30–32	0.99	200	6.7
240858	2007-05-12 00:48:27	93.62	9–31	0.97	209	6.6
*241489	2007-05-13 00:41:41	94.07	8–29	0.92	201	5.9
*241979	2007-05-14 00:37:41	94.07	8–30	0.94	201	6.4
242582	2007-05-15 00:38:26	93.55	9–30	0.94	207	6.4
243215	2007-05-16 00:37:12	93.92	10–32	0.95	206	6.2
244362	2007-05-18 00:20:57	94.43	8–30	0.97	191	5.3
244883	2007-05-19 00:19:27	55.95	9–22	0.96	197	5.5
245461	2007-05-20 00:14:57	36.48	8–17	0.90	202	5.4
338344	2008-03-02 03:33:10	24.18	11–17	0.92	161	6.2
338660	2008-03-04 03:04:24	3.78	20–21	0.92	162	6.4
338668	2008-03-04 03:24:40	47.15	6–16	0.93	166	5.7
338711	2008-03-04 04:22:26	22.18	3–4	0.93	161	4.8
339404	2008-03-06 04:49:55	21.95	6–9	0.94	157	4.8
339773	2008-03-07 02:23:28	57.73	15–28	0.95	197	6.6
339830	2008-03-07 03:26:10	126.83	3–16	0.93	202	5.8
340316	2008-03-08 02:22:24	37.78	19–27	0.97	177	5.7
340352	2008-03-08 03:03:39	144.55	3–18	0.95	189	5.7
340857	2008-03-09 02:20:13	96.88	6–27	0.93	201	5.8
340954	2008-03-09 04:03:25	18.45	3–4	0.90	207	5.2
340975	2008-03-09 04:29:40	59.75	4–16	0.91	205	4.7
341262	2008-03-10 02:12:25	93.82	7–28	0.93	195	6.7

Table A.16: Data sequences of 2E 1415.6+2557.

A Data compendium

2E 1415.6+2557 continued

Sequence	Time	Duration [min]	Zd [°]	Ped. RMS	Data rate [Hz]	Inhom.
341655	2008-03-11 02:10:24	185.43	3–28	0.93	216	6.9
344202	2008-03-29 01:26:41	37.23	13–22	0.95	204	8.5
344635	2008-03-30 01:30:10	25.02	13–19	0.89	192	8.1
345124	2008-04-01 01:32:27	19.75	13–18	0.94	204	8.7
345587	2008-04-02 01:15:13	33.22	13–21	0.91	209	9.0
346642	2008-04-05 01:17:13	23.87	12–18	0.90	213	8.4

PKS 1424+240

Sequence	Time	Duration [min]	Zd [°]	Ped. RMS	Data rate [Hz]	Inhom.
91579	2006-05-23 23:36:41	58.18	5–15	0.95	233	6.6
*91715	2006-05-25 00:07:11	39.85	10–20	0.96	217	10.0
91999	2006-05-25 23:22:50	77.87	5–18	0.93	230	6.2
*92123	2006-05-26 23:22:05	81.58	5–19	0.94	221	8.5
222047	2007-02-20 05:12:39	72.03	5–13	0.89	212	5.8
*222690	2007-02-21 02:57:33	194.82	5–35	0.88	228	6.8
*223226	2007-02-22 02:51:53	189.20	5–36	0.89	218	6.8
223948	2007-02-23 03:35:09	146.83	5–25	0.87	222	6.6
*224387	2007-02-24 02:55:53	162.33	5–34	0.83	215	6.7
*224705	2007-02-25 02:58:23	187.05	5–32	0.86	213	7.1

RX J1725.0+1152

Sequence	Time	Duration [min]	Zd [°]	Ped. RMS	Data rate [Hz]	Inhom.
264216	2007-08-09 22:38:16	32.40	24–30	0.88	207	8.5
264823	2007-08-10 22:34:46	32.17	24–30	0.90	207	6.9
265231	2007-08-11 22:23:32	36.68	23–30	0.89	192	7.2
265862	2007-08-12 22:27:47	32.52	25–30	0.95	201	7.4
266566	2007-08-13 22:18:30	36.68	23–31	0.88	203	7.2
267253	2007-08-14 22:33:31	22.45	27–31	0.88	207	5.6
267929	2007-08-15 22:13:15	33.07	24–30	0.91	179	4.9
268515	2007-08-16 21:08:50	1.17	17–17	0.97	237	7.3
268519	2007-08-16 21:39:01	6.15	20–20	0.91	183	9.2

Table A.17: Data sequences of 2E 1415.6+2557, PKS 1424+240 and RX J1725.0+1152.

RX J1725.0+1152 continued

Sequence	Time	Duration [min]	Zd [°]	Ped. RMS	Data rate [Hz]	Inhom.
268530	2007-08-16 21:48:01	55.90	21–31	0.89	224	8.4
270174	2007-08-18 21:17:01	19.48	18–20	1.17	205	6.9
270241	2007-08-18 21:58:30	34.85	23–30	1.03	201	7.2
270918	2007-08-19 21:11:16	75.22	18–30	1.33	171	6.3
271720	2007-08-20 21:36:17	46.52	21–29	1.65	124	8.9
274297	2007-08-31 21:12:16	26.75	24–30	1.01	210	5.8
274363	2007-09-01 20:52:01	27.82	22–26	0.86	227	6.2
274415	2007-09-01 21:34:30	6.87	29–30	0.86	226	6.3
274549	2007-09-02 20:47:30	44.15	21–30	0.89	197	7.1
275086	2007-09-03 20:51:16	37.83	22–30	0.91	190	6.5
275788	2007-09-04 21:16:16	18.28	28–32	0.90	210	7.4
276218	2007-09-05 21:00:33	29.78	26–31	0.89	217	7.1
277148	2007-09-07 20:49:02	31.22	25–30	0.91	216	7.8
278160	2007-09-09 20:34:45	5.88	24–25	0.96	228	6.5
278170	2007-09-09 20:43:01	25.10	26–30	0.92	226	6.5
278700	2007-09-10 20:36:46	33.30	25–32	0.92	216	7.9
339106	2008-03-05 05:51:41	13.93	29–31	1.02	114	6.7
339955	2008-03-07 05:42:41	29.88	26–31	1.01	192	5.3
340489	2008-03-08 05:38:40	31.35	26–31	1.03	183	5.7
341037	2008-03-09 05:37:55	33.82	24–31	1.00	197	5.1
341847	2008-03-11 05:28:40	37.95	24–30	0.99	207	7.0
342332	2008-03-12 05:20:39	48.83	22–31	1.17	157	4.7
342684	2008-03-13 05:18:27	51.30	22–31	1.07	132	5.1
346814	2008-04-05 04:20:42	72.15	17–25	0.97	208	8.4
347358	2008-04-12 05:01:13	25.35	17–17	1.03	194	9.3
347686	2008-04-15 03:58:57	31.53	18–22	0.98	198	8.7
347732	2008-04-15 05:09:28	15.08	17–18	0.97	205	8.0
347830	2008-04-17 04:57:27	25.33	17–18	1.04	198	7.5
1005362	2009-03-30 03:53:13	104.20	18–35	1.05	201	11.3
1005409	2009-03-31 04:48:43	53.28	17–24	1.07	202	12.2
1005455	2009-04-01 05:01:57	42.75	17–21	1.08	185	23.0
1005499	2009-04-02 04:34:12	69.33	17–25	1.07	198	11.9
1005544	2009-04-03 05:02:57	9.68	19–20	1.03	199	13.1
1005594	2009-04-04 04:33:58	66.52	17–24	1.05	275	8.9
1005606	2009-04-05 04:13:28	81.30	17–26	1.04	191	13.0
1005619	2009-04-06 04:47:28	52.72	17–21	1.04	200	13.6

Table A.18: Data sequences of RX J1725.0+1152.

A Data compendium

1ES 1727+502

Sequence	Time	Duration [min]	Zd [°]	Ped. RMS	Data rate [Hz]	Inhom.
92493	2006-06-01 02:03:44	62.48	21–25	0.98	183	17.0
92586	2006-06-02 02:12:05	58.22	22–25	0.98	174	19.6
*92705	2006-06-03 02:01:47	30.93	22–23	1.01	179	17.7
92781	2006-06-04 01:55:44	74.10	22–26	0.98	167	21.1
92840	2006-06-05 02:11:32	54.00	22–26	0.97	176	19.1
*238264	2007-04-27 04:17:57	49.48	21–23	0.92	162	6.9
239742	2007-05-10 01:01:41	60.10	26–36	0.99	161	7.3

1ES 1741+196

Sequence	Time	Duration [min]	Zd [°]	Ped. RMS	Data rate [Hz]	Inhom.
96662	2006-07-27 22:10:01	38.25	9–10	0.99	227	4.4
97655	2006-08-12 21:50:04	53.83	10–19	1.43	180	5.9
97673	2006-08-13 21:03:39	25.95	9–10	1.04	186	5.3
97682	2006-08-13 21:33:16	18.62	10–11	0.99	179	6.4
97689	2006-08-13 21:53:40	18.65	11–14	1.01	173	7.7
97699	2006-08-13 22:19:55	18.67	16–19	1.02	175	7.6
97707	2006-08-13 22:40:19	18.63	20–22	1.26	177	7.9
97714	2006-08-13 23:00:37	10.50	24–26	1.54	168	7.1
97722	2006-08-14 21:07:42	18.05	9–10	1.04	186	6.4
*97729	2006-08-14 21:28:34	18.70	9–10	1.03	172	9.1
97736	2006-08-14 21:49:01	18.65	12–14	1.02	172	8.7
*97743	2006-08-14 22:09:25	18.65	14–17	1.04	150	8.7
*97751	2006-08-14 22:29:49	18.65	19–22	1.02	145	9.1
*97765	2006-08-14 23:10:32	18.60	28–30	1.15	138	9.2
*97775	2006-08-14 23:51:58	21.15	36–40	1.62	115	7.4
*232410	2007-04-14 04:40:25	41.58	9–14	1.00	199	6.4
*233133	2007-04-15 04:40:25	27.20	10–13	0.92	212	6.4
233832	2007-04-16 04:44:55	35.55	9–12	0.97	205	7.5
*234032	2007-04-17 04:30:54	14.93	11–13	0.93	216	6.1
235917	2007-04-20 04:22:57	50.75	9–13	0.91	211	6.6

Table A.19: Data sequences of 1ES 1727+502 and 1ES 1741+196.

1ES 1741+196 continued

Sequence	Time	Duration [min]	Zd [°]	Ped. RMS	Data rate [Hz]	Inhom.
236594	2007-04-21 04:24:58	47.85	9–12	1.03	194	7.1
237677	2007-04-23 04:15:57	55.82	9–12	0.93	206	6.0
237923	2007-04-24 04:07:42	61.22	9–13	0.95	201	6.4
238064	2007-04-25 02:56:28	56.07	14–26	0.97	181	6.8
238145	2007-04-25 03:58:12	69.43	9–14	0.95	193	6.7

B3 2247+381

Sequence	Time	Duration [min]	Zd [°]	Ped. RMS	Data rate [Hz]	Inhom.
97359	2006-08-03 04:46:05	21.12	22–26	1.04	173	4.2
97451	2006-08-04 04:45:20	38.90	23–30	1.12	163	3.7
*97553	2006-08-05 04:41:26	38.57	22–30	1.10	214	3.6
97604	2006-08-07 04:00:14	72.83	17–29	1.06	162	4.4
98531	2006-08-22 03:34:18	62.65	22–34	1.04	194	4.7
98674	2006-08-23 03:27:22	66.95	22–34	1.03	189	6.1
99604	2006-09-01 01:23:50	99.12	10–24	1.03	177	4.2
99715	2006-09-02 00:48:33	26.13	10–11	1.06	173	6.2
99725	2006-09-02 01:19:48	81.95	10–21	1.02	175	5.4

Table A.20: Data sequences of 1ES 1741+196 and B3 2247+381.

A Data compendium

Off 3EG 1835+5918

Sequence	Time	Duration [min]	Zd [°]	Ped. RMS	Data rate [Hz]	Inhom.
94520	2006-06-27 02:26:26	94.70	34–43	0.98	151	22.1
94576	2006-06-27 22:16:31	35.80	38–41	0.96	145	22.2
95066	2006-07-02 00:24:19	120.03	31–35	0.98	166	14.6
95177	2006-07-03 02:22:28	76.32	36–43	0.98	168	10.5

Off HB89 1721+343

Sequence	Time	Duration [min]	Zd [°]	Ped. RMS	Data rate [Hz]	Inhom.
95618	2006-07-17 23:15:47	33.38	6–10	0.94	161	6.0
95631	2006-07-18 00:03:29	34.83	14–20	0.95	163	6.5
96262	2006-07-23 22:53:26	60.42	6–16	0.99	124	6.2
96277	2006-07-23 23:55:41	60.40	17–29	1.02	113	6.1

Off GRB (1)

Sequence	Time	Duration [min]	Zd [°]	Ped. RMS	Data rate [Hz]	Inhom.
112092	2007-01-21 21:20:34	201.63	49–56	1.09	140	8.3

Off GRB (2)

Sequence	Time	Duration [min]	Zd [°]	Ped. RMS	Data rate [Hz]	Inhom.
232609	2007-04-14 22:29:53	48.62	12–17	0.87	209	6.4
232688	2007-04-14 23:22:39	61.97	11–16	0.89	209	6.1
232788	2007-04-15 00:29:25	33.97	16–22	0.89	205	6.5
232844	2007-04-15 01:06:40	22.18	23–27	0.90	204	6.2

Off HB89 0954+658

Sequence	Time	Duration [min]	Zd [°]	Ped. RMS	Data rate [Hz]	Inhom.
106669	2006-11-30 03:53:05	58.57	42–47	1.05	162	8.2
106686	2006-11-30 04:53:53	58.20	38–42	1.05	168	8.8

Table A.21: Data sequences of the crosscheck sample. Used positions: Off pointings to 3EG 1835+5918, HB89 1721+343, GRBs (two different pointings) and HB89 0954+658.

Off 3C 273

Sequence	Time	Duration [min]	Zd [°]	Ped. RMS	Data rate [Hz]	Inhom.
240089	2007-05-10 22:46:11	75.60	28–37	1.00	173	7.1

Willman 1

Sequence	Time	Duration [min]	Zd [°]	Ped. RMS	Data rate [Hz]	Inhom.
329294	2008-01-17 02:52:22	28.18	25–27	0.94	195	8.9
334000	2008-02-05 01:11:24	62.20	24–30	0.89	210	7.6
334125	2008-02-05 03:16:24	64.45	22–27	0.87	215	6.8
335115	2008-02-07 01:17:23	41.27	25–29	0.86	179	8.2
335160	2008-02-07 03:29:08	27.45	23–25	0.84	189	7.6
335518	2008-02-08 01:00:37	62.87	24–30	0.95	168	6.6
335627	2008-02-08 03:05:54	32.40	22–27	0.87	191	6.8
343758	2008-03-27 23:56:25	61.63	22–27	0.90	201	7.7
344030	2008-03-28 22:17:55	58.75	22–26	0.92	197	7.7
344144	2008-03-29 00:20:54	59.83	24–30	0.90	195	7.2
344473	2008-03-29 22:11:25	59.90	22–27	0.91	185	7.9
344578	2008-03-30 00:16:25	57.13	24–30	0.89	180	7.4
345010	2008-03-30 23:57:57	75.15	23–30	0.88	196	7.5
345506	2008-04-01 23:48:13	53.87	23–27	0.92	203	7.2
345577	2008-04-02 01:02:42	7.70	30–31	0.92	199	7.4
345913	2008-04-02 22:02:14	53.67	22–26	0.93	175	7.2
346016	2008-04-03 00:05:58	54.25	24–30	0.94	159	7.4
346332	2008-04-03 22:23:12	27.38	22–24	0.88	204	7.7
346425	2008-04-04 00:04:13	10.87	24–26	0.88	197	7.0
346523	2008-04-04 21:55:57	51.62	22–26	0.88	206	7.9
346971	2008-04-05 21:52:12	51.43	22–26	0.92	202	7.9
347090	2008-04-05 23:57:13	2.75	25–25	0.92	190	7.2
348316	2008-04-27 23:03:26	68.22	28–38	1.00	133	8.7

Table A.22: Data sequences of the crosscheck sample. Used positions: Off pointing to 3C 273 and On pointing to Willman 1.

A Data compendium

Off Willman 1

Sequence	Time	Duration [min]	Zd [°]	Ped. RMS	Data rate [Hz]	Inhom.
334066	2008-02-05 02:17:38	54.55	22–26	0.87	213	7.4
335155	2008-02-07 02:17:37	2.03	25–26	0.84	182	8.6
335574	2008-02-08 02:06:37	55.88	22–26	0.88	186	6.9
343708	2008-03-27 23:03:25	48.73	22–25	0.89	205	8.8
344089	2008-03-28 23:20:39	55.60	22–24	0.90	198	7.5
344529	2008-03-29 23:16:10	52.50	22–24	0.89	185	7.8
344954	2008-03-30 22:55:58	57.13	22–25	0.88	198	8.8
345456	2008-04-01 22:55:13	49.00	22–24	0.91	205	8.0
345970	2008-04-02 23:10:12	43.08	22–23	0.93	168	7.3
346369	2008-04-03 23:04:27	46.23	22–23	0.87	206	7.7
346584	2008-04-04 22:58:42	48.30	22–23	0.88	205	7.3
347036	2008-04-05 22:59:57	43.98	22–23	0.92	202	7.5

Draco

Sequence	Time	Duration [min]	Zd [°]	Ped. RMS	Data rate [Hz]	Inhom.
239552	2007-05-09 01:30:43	11.47	34–36	1.11	117	6.6
239819	2007-05-10 02:08:27	53.83	29–32	1.16	160	6.3
239886	2007-05-10 03:09:11	19.17	29–29	1.40	134	8.1
240383	2007-05-11 02:43:57	17.63	29–30	1.07	191	6.5
240413	2007-05-11 03:05:57	17.67	29–29	1.16	174	5.8
241643	2007-05-13 02:28:42	0.75	30–30	0.92	190	6.7
241647	2007-05-13 02:41:55	17.67	29–29	0.94	187	5.7
242130	2007-05-14 02:24:57	56.13	29–30	0.99	164	6.2
242738	2007-05-15 02:25:43	36.85	29–30	0.95	197	6.5
244503	2007-05-18 02:07:42	56.37	29–30	1.03	172	5.5
244970	2007-05-19 01:26:11	94.32	29–32	0.98	180	5.9
245521	2007-05-20 01:00:42	56.00	30–34	0.94	185	5.5

Table A.23: Data sequences of the crosscheck sample. Used positions: Off pointing to Willman 1 and On pointing to Draco.

B ϑ^2-distributions

In this chapter a list of all ϑ^2-distributions of the three Crab Nebula datasets and the 20 BL Lac objects in order of increasing rightascension as well as of the crosscheck sample are given. Shown are the On events as black crosses, the Off events as grey shaded area and the ϑ^2-cut as dashed vertical line.

B ϑ^2-distributions

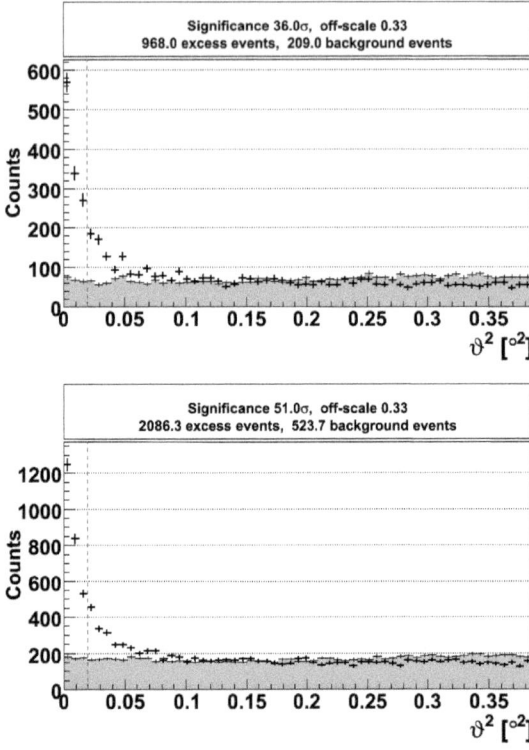

Figure B.1: ϑ^2-distributions of the Crab Nebula datasets. Upper panel: 300 MHz FADCs w/o optical splitters; Lower Panel: 300 MHz FADCs w/ optical splitters.

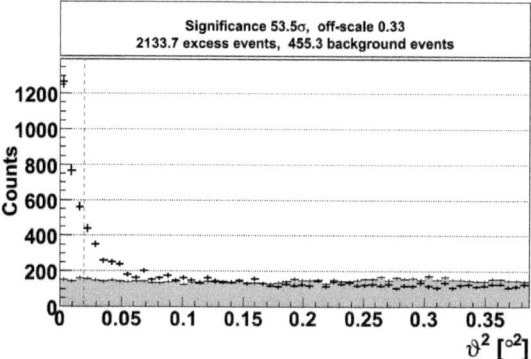

Figure B.2: ϑ^2-distribution of the Crab Nebula datasets for the 2 GHz MUX FADCs.

B ϑ^2-distributions

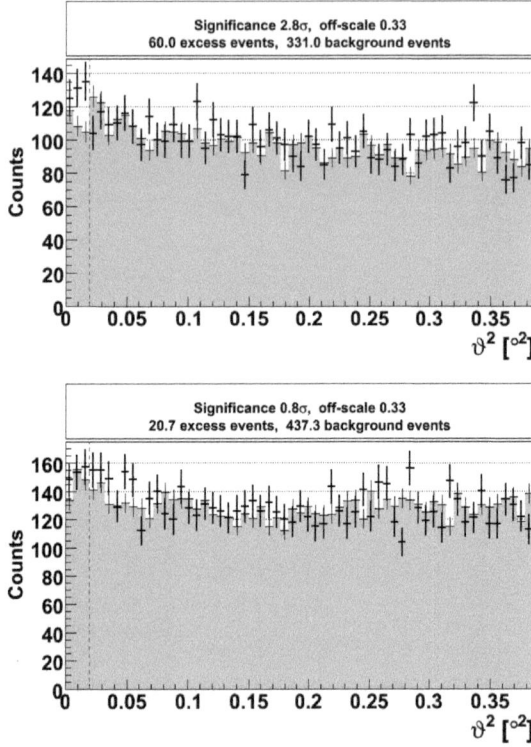

Figure B.3: ϑ^2-distributions of the BL Lac object sample. Upper panel: 1ES 0033+595; Lower panel: 1ES 0120+340.

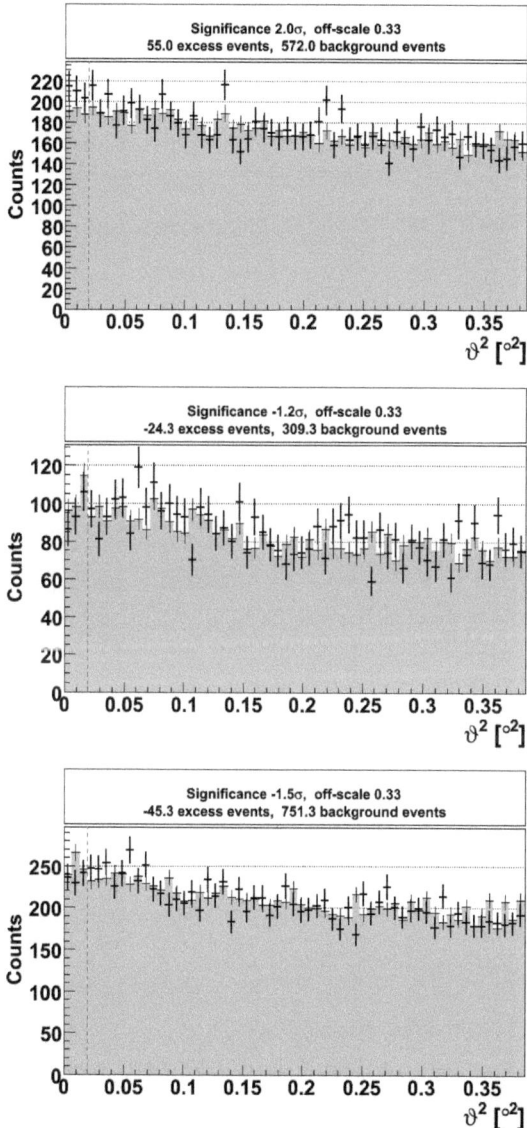

Figure B.4: ϑ^2-distributions of the BL Lac object sample. Upper panel: 1ES 0229+200; Mid panel: RX J0319.8+1845; Lower panel: 1ES 0323+022.

B ϑ^2-distributions

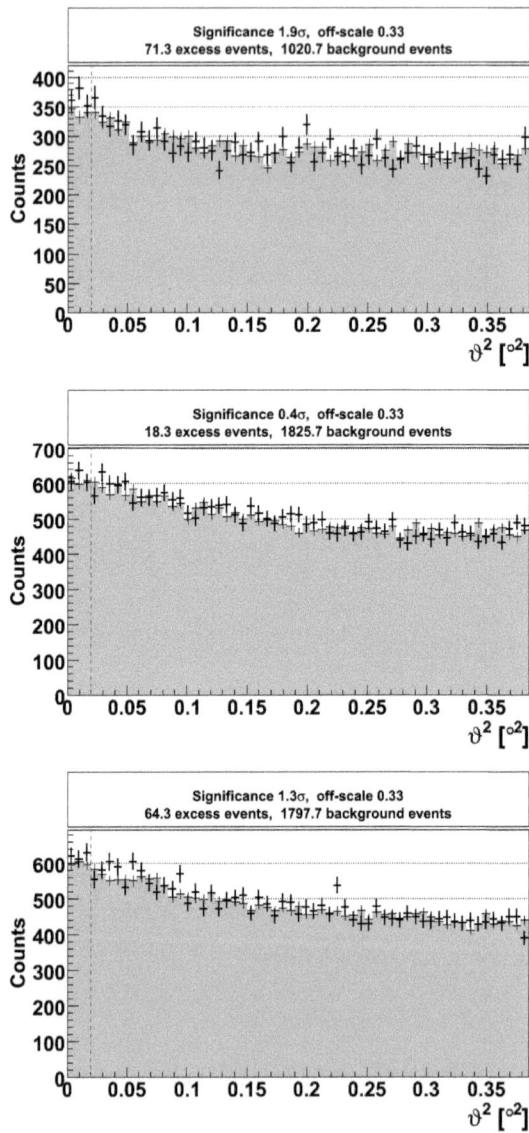

Figure B.5: ϑ^2-distributions of the BL Lac object sample. Upper panel: 1ES 0414+009; Mid panel: 1RXS J044127.8+150455; Lower panel: 1ES 0647+250.

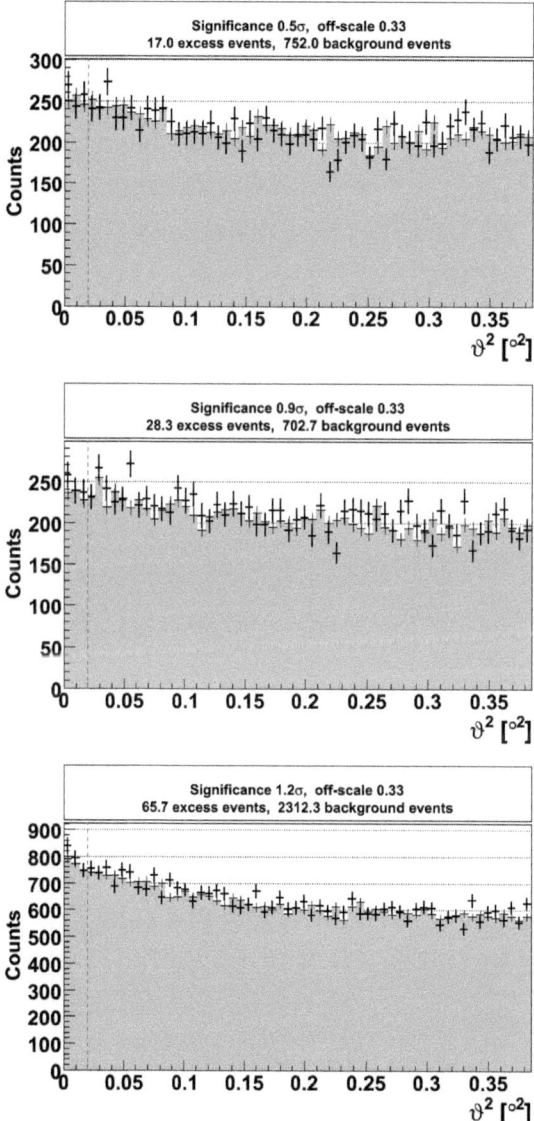

Figure B.6: ϑ^2-distributions of the BL Lac object sample. Upper panel: 1ES 0806+524; Mid panel: 1ES 0927+500; Lower panel: 1ES 1028+511.

B ϑ^2-distributions

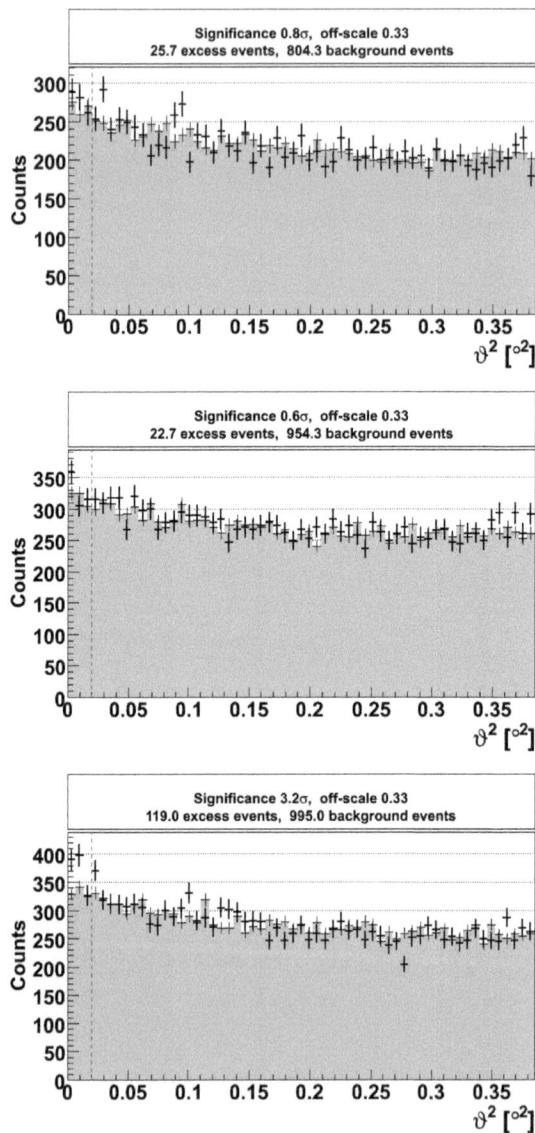

Figure B.7: ϑ^2-distributions of the BL Lac object sample. Upper panel: RGB J1117+202; Mid panel: RXS J1136.5+6737; Lower panel: B2 1215+30.

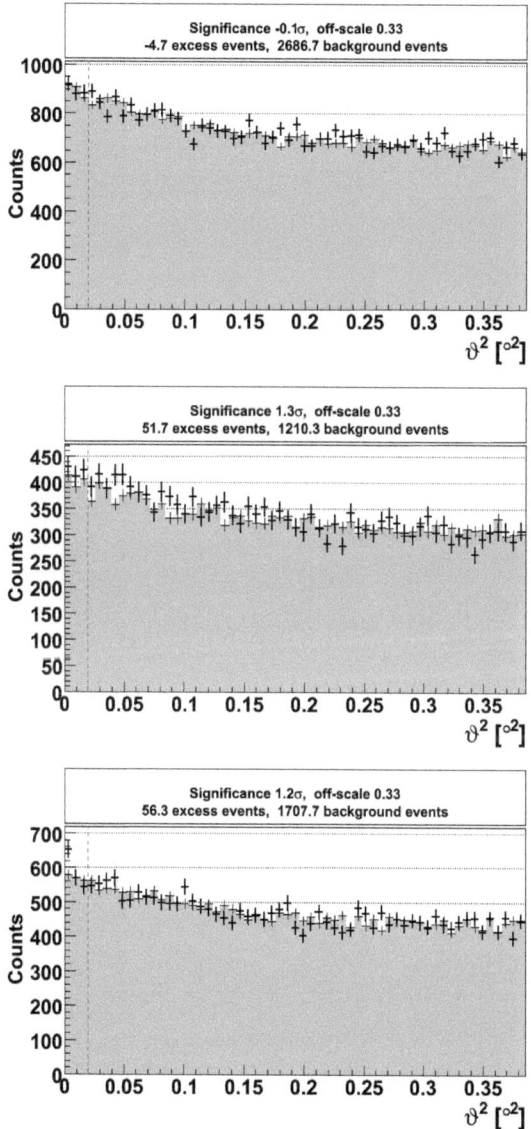

Figure B.8: ϑ^2-distributions of the BL Lac object sample. Upper panel: 2E 1415.6+2557; Mid panel: PKS 1424+240; Lower panel: RX J1725.0+1152.

B ϑ^2-distributions

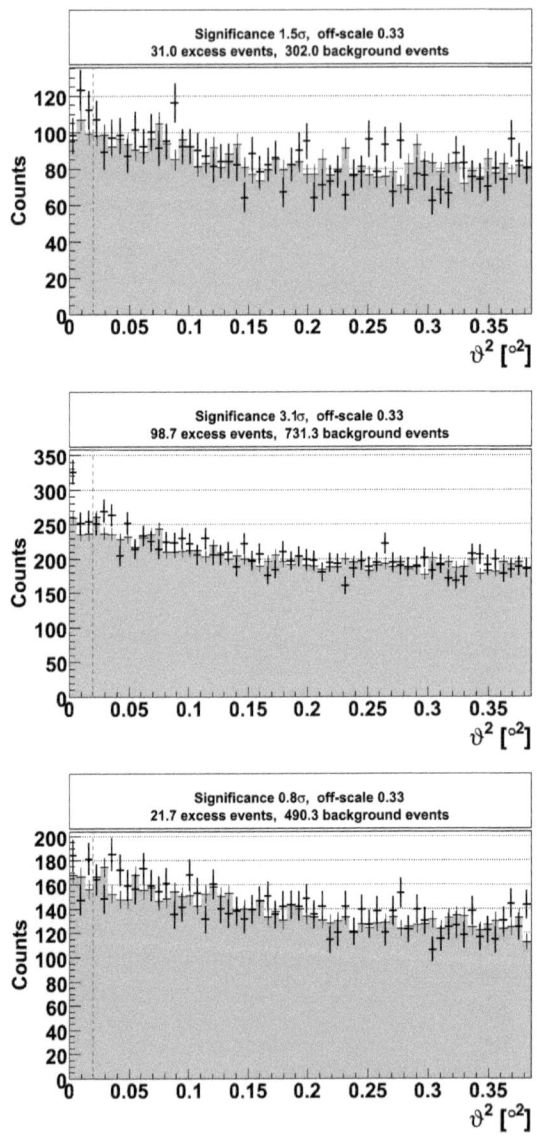

Figure B.9: ϑ^2-distributions of the BL Lac object sample. Upper panel: 1ES 1727+502; Mid panel: 1ES 1741+196; Lower panel: B3 2247+381.

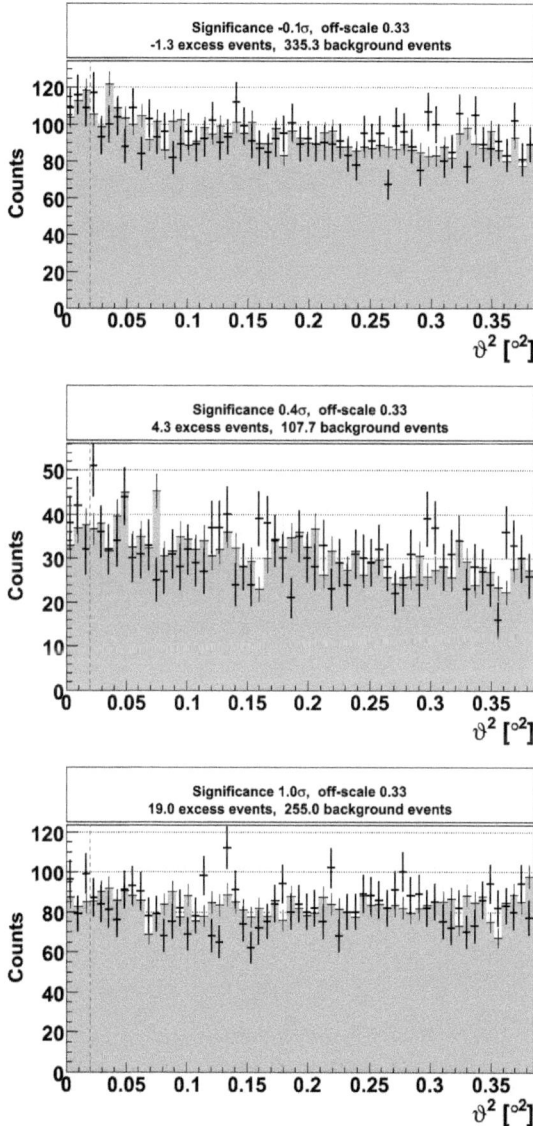

Figure B.10: ϑ^2-distributions of the crosscheck sample.

B ϑ^2-distributions

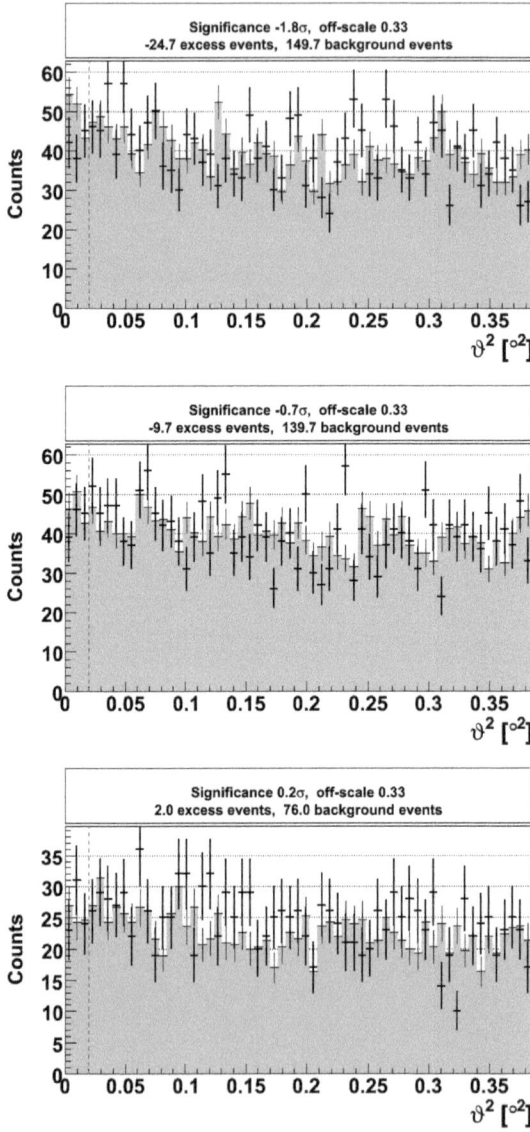

Figure B.11: ϑ^2-distributions of the crosscheck sample, continued.

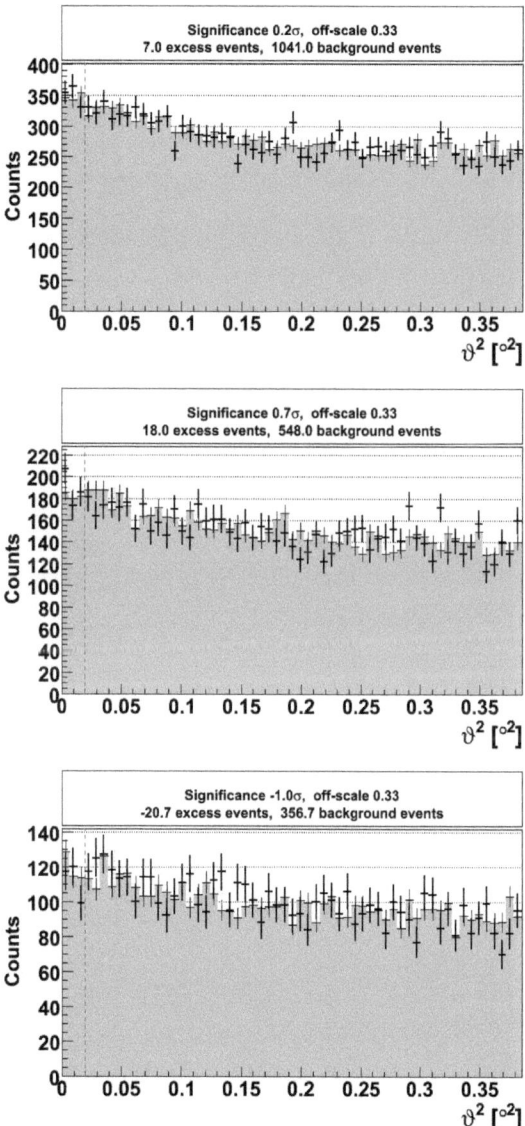

Figure B.12: ϑ^2-distributions of the crosscheck sample, continued.

B ϑ^2-*distributions*

C Lightcurves

The lightcurves presented here are based on excess and background event rates. None of the individual objects reveals any significant variability. Shown are the excess event rates (blue marks) and the background event rates (red marks) with the corresponding mean values as dashed lines.

C Lightcurves

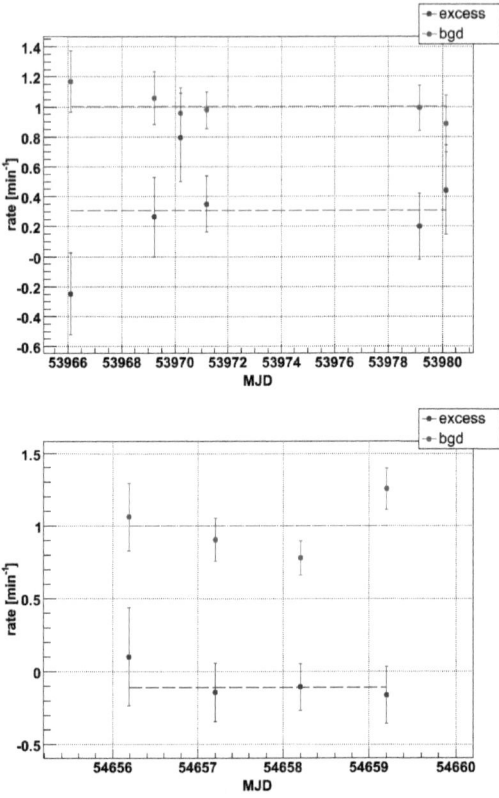

Figure C.1: Lightcurves of 1ES 0033+595 in 2006 and 2008.

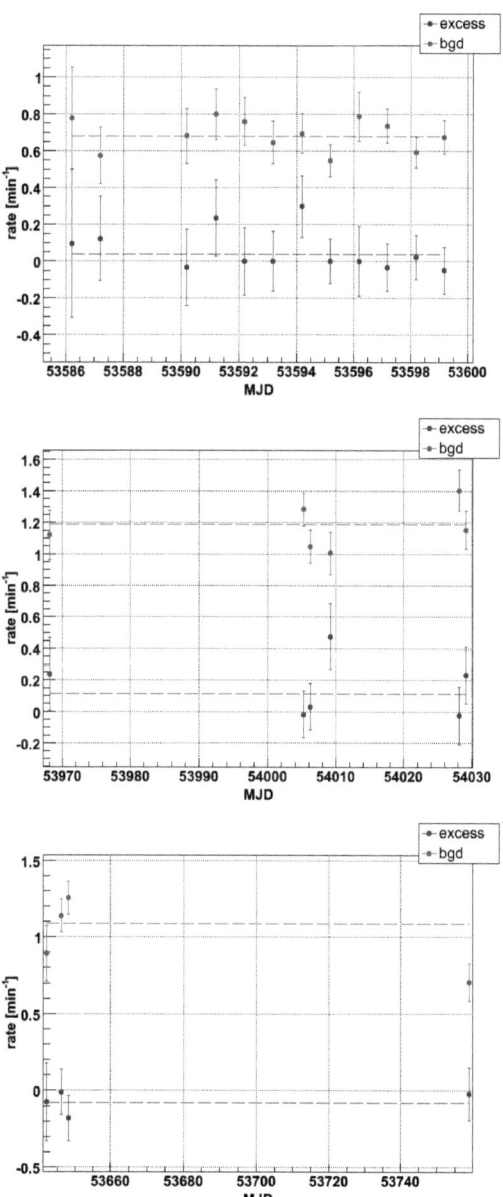

Figure C.2: Lightcurves of the BL Lac object sample. Upper panel: 1ES 0120+340; Mid panel: 1ES0229+200; Lower panel: RX J0319.8+1845.

C Lightcurves

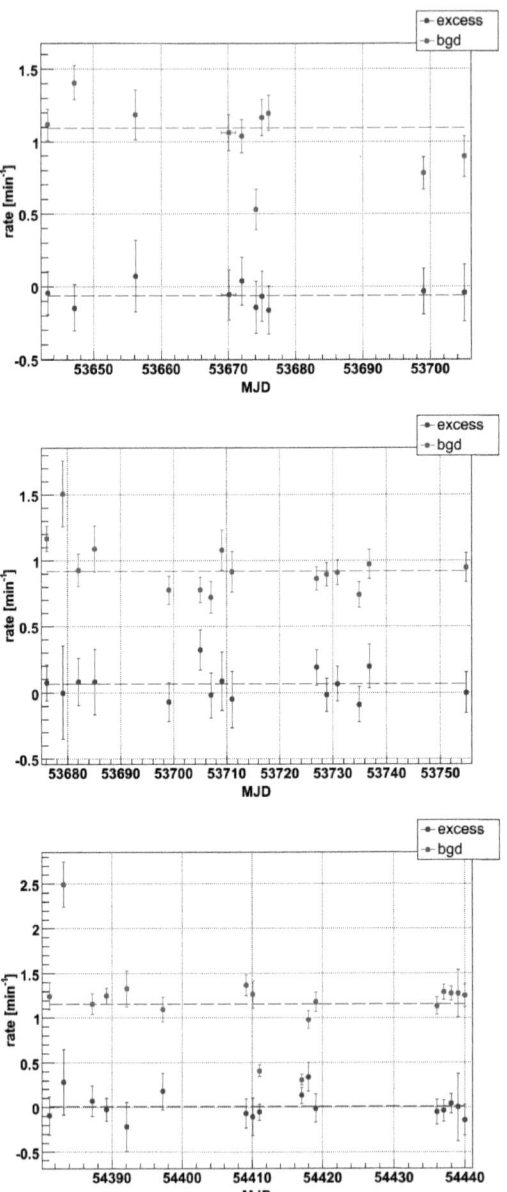

Figure C.3: Lightcurves of the BL Lac object sample. Upper panel: 1ES 0323+022; Mid panel: 1ES0414+009; Lower panel: 1RXS J044127.8+150455.

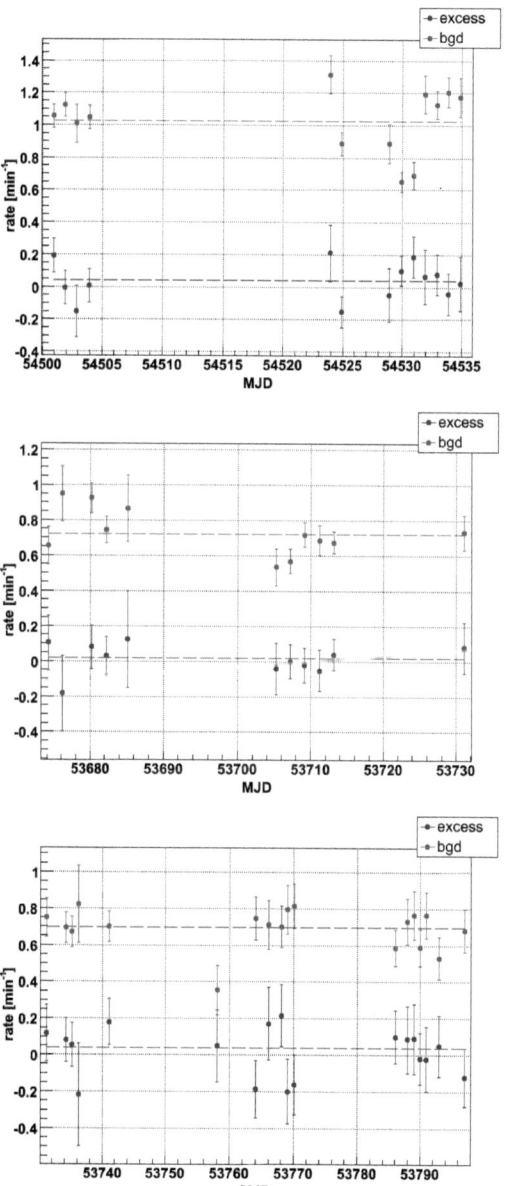

Figure C.4: Lightcurves of the BL Lac object sample. Upper panel: 1ES0647+250; Mid panel: 1ES0806+524; Lower panel: 1ES 0927+500.

C Lightcurves

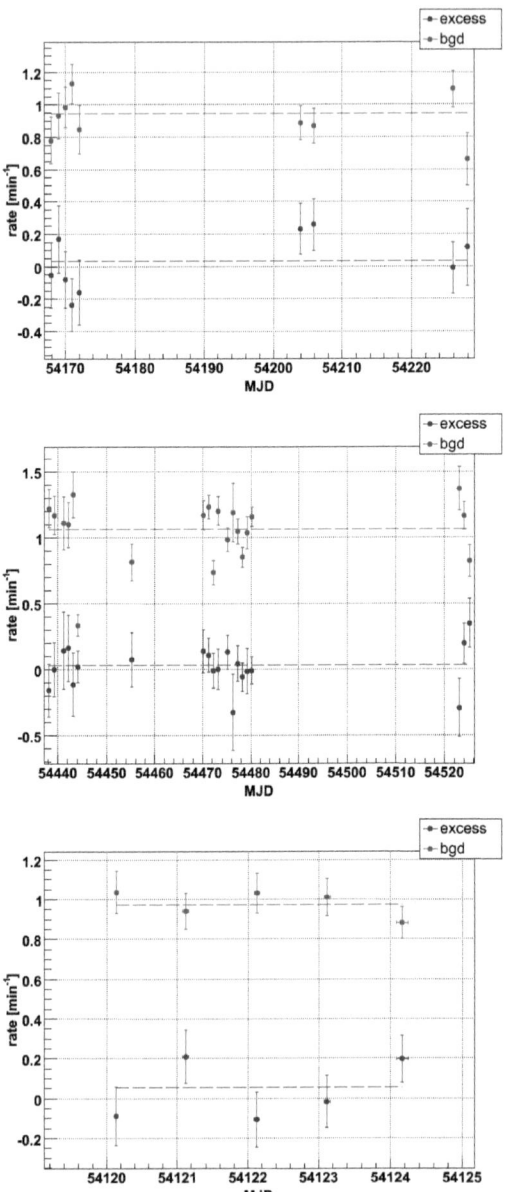

Figure C.5: Lightcurves of the BL Lac object sample. Upper and mid panel: 1ES1028+511 in 2007 and 2008; Lower panel: RGB J1117+202 in 2007.

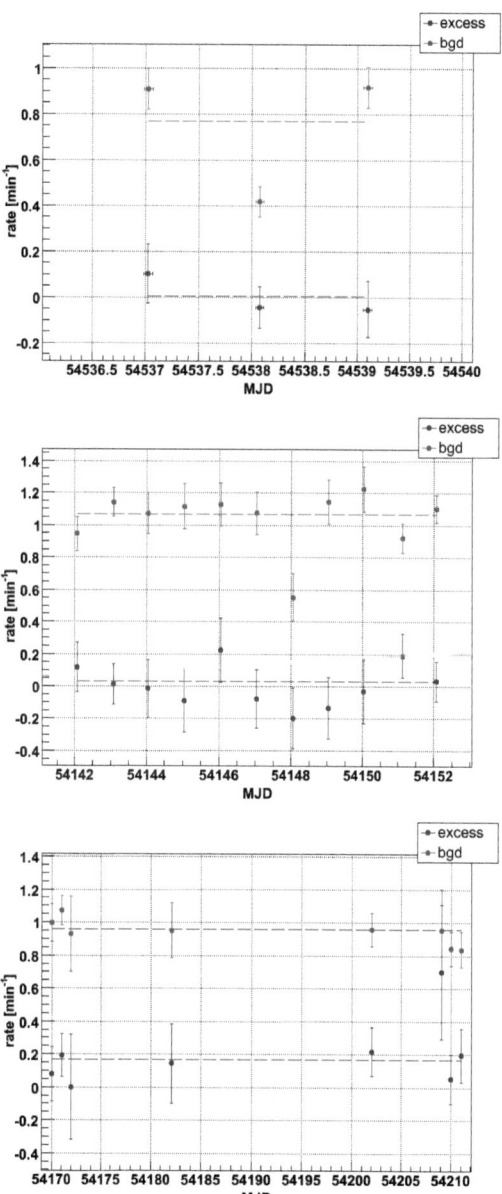

Figure C.6: Lightcurves of the BL Lac object sample. Upper panel: RGB J1117+202 in 2008; Mid panel: RXS J1136.5+6737; Lower panel: B2 1215+30 in 2007.

C Lightcurves

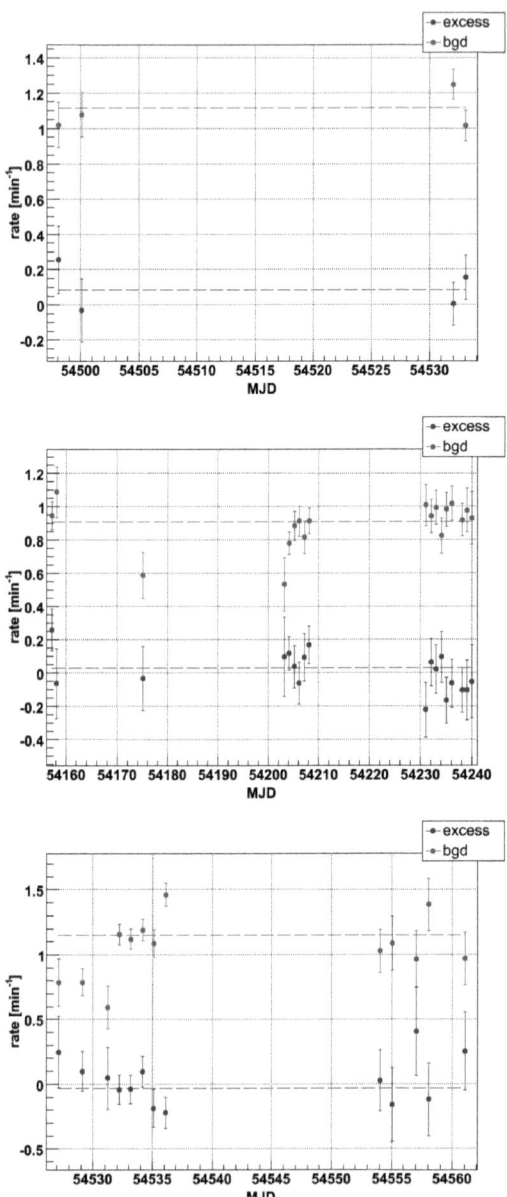

Figure C.7: Lightcurves of the BL Lac object sample. Upper panel: B2 1215+30 in 2008; Mid and lower panel: 2E 1415.6+2557 in 2007 and 2008.

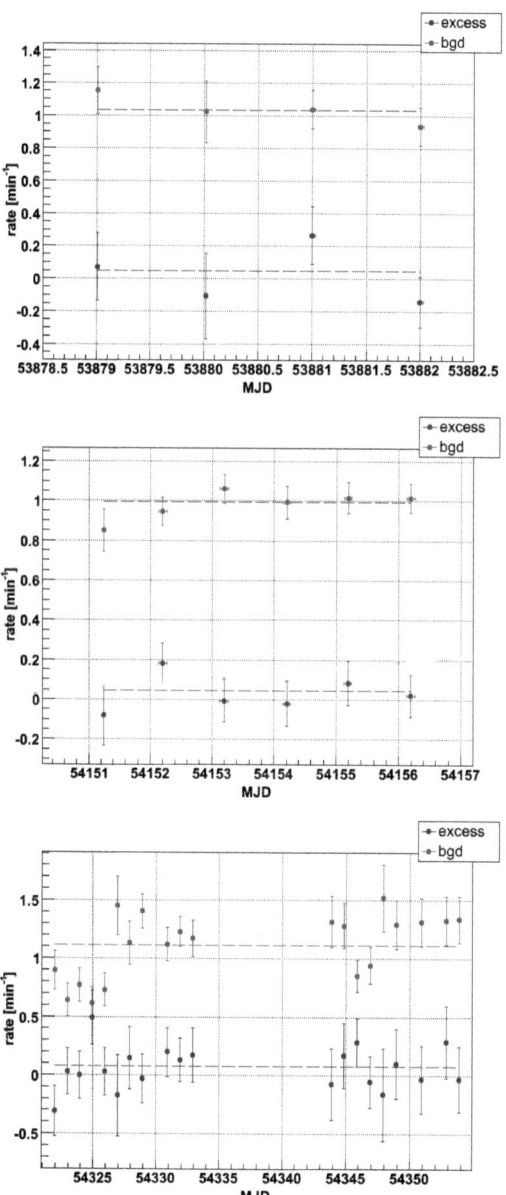

Figure C.8: Lightcurves of the BL Lac object sample. Upper and mid panel: PKS 1424+240 in 2006 and 2007; Lower panel: RX J1725.0+1152 in 2007.

C Lightcurves

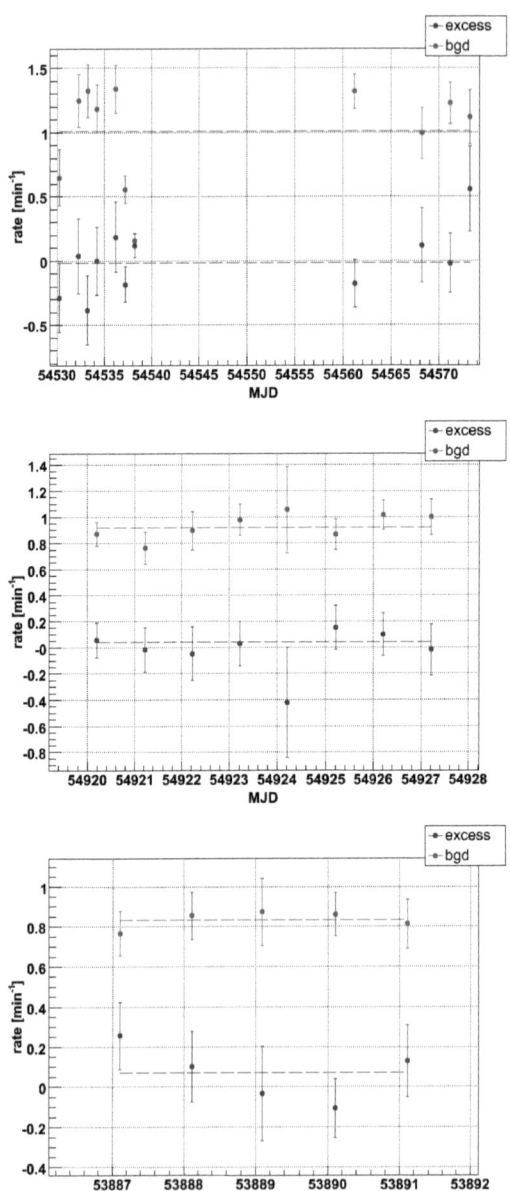

Figure C.9: Lightcurves of the BL Lac object sample. Upper and mid panel: RX J1725.0+1152 in 2008 and 2009; Lower panel: 1ES 1727+502 in 2006.

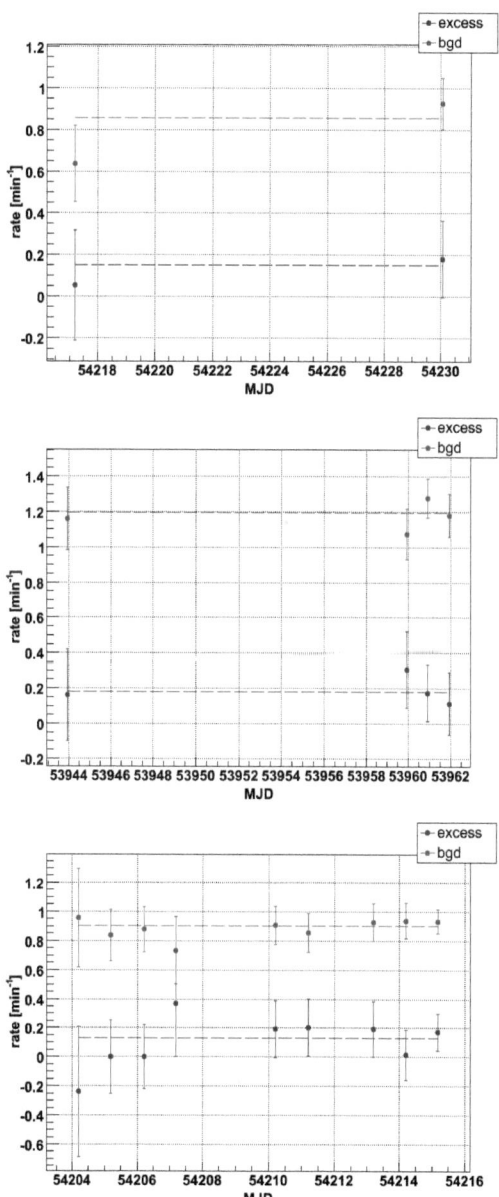

Figure C.10: Lightcurves of the BL Lac object sample. Upper panel: 1ES 1727+502 2007; Mid and lower panel: 1ES1741+196 in 2006 and 2007.

C Lightcurves

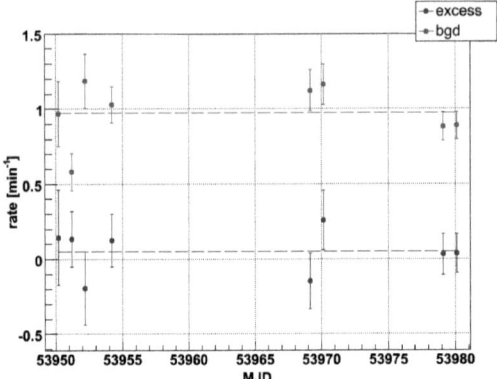

Figure C.11: Lightcurve of B3 2247+381.

List of Figures

1.1 AGN paradigm. 14
1.2 Exemplary spectral energy distribution 15
1.3 Spectral blazar sequence 17

2.1 Cosmic ray spectrum 25
2.2 Hillas plot 26
2.3 GRB scenario 30
2.4 M82 31
2.5 Artist view of a magnetar 32
2.6 Artist view of an AGN 33

3.1 Opacity of the atmosphere for photons 36
3.2 CORSIKA simulation of a 100 GeV γ- and proton shower 37
3.3 The MAGIC telescope 40

4.1 Calibrated and cleaned image of a γ-like shower in the MAGIC camera. 49
4.2 ϑ^2-distribution of the combined Crab Nebula data sample. 57
4.3 Differential energy spectra of the Crab Nebula. 58

5.1 ϑ^2-plot for 1ES 1028+511. 74
5.2 Comparison of *area* vs *size* of data and MC simulations. 74
5.3 Lightcurve for RXS J1136.5+6737. 75
5.4 Significance distribution of the blazar and the crosscheck samples. 77
5.5 ϑ^2-distribution of the stacked blazar sample. 79
5.6 VHE spectrum obtained with the stacking method. 80
5.7 Cumulative excess events and significance versus t_{eff}. 81
5.8 Stacked ϑ^2-distribution of the crosscheck sample. 83

List of Figures

6.1	Deabsorbed differential energy spectrum	89
6.2	Broad-band spectral indices $\alpha_{X-\gamma}$ vs X-ray luminosity.	91
6.3	Spectral energy distribution of the stacked BL Lac object sample.	93
6.4	Broad-band spectral indices $\alpha_{X-\gamma}$ vs. X-ray luminosity including measured low-state sources.	96
6.5	SED of the stacked BL Lac object sample and measured low state sources.	97
B.1	ϑ^2-distributions of the Crab Nebula datasets (1).	124
B.2	ϑ^2-distributions of the Crab Nebula datasets (2).	125
B.3	ϑ^2-distributions of the BL Lac object sample (1).	126
B.4	ϑ^2-distributions of the BL Lac object sample (2).	127
B.5	ϑ^2-distributions of the BL Lac object sample (3).	128
B.6	ϑ^2-distributions of the BL Lac object sample (4).	129
B.7	ϑ^2-distributions of the BL Lac object sample (5).	130
B.8	ϑ^2-distributions of the BL Lac object sample (6).	131
B.9	ϑ^2-distributions of the BL Lac object sample (7).	132
B.10	ϑ^2-distributions of the crosscheck sample (1).	133
B.11	ϑ^2-distributions of the crosscheck sample (2).	134
B.12	ϑ^2-distributions of the crosscheck sample (3).	135
C.1	Lightcurves of the BL Lac object sample (1).	138
C.2	Lightcurves of the BL Lac object sample (2).	139
C.3	Lightcurves of the BL Lac object sample (3).	140
C.4	Lightcurves of the BL Lac object sample (4).	141
C.5	Lightcurves of the BL Lac object sample (5).	142
C.6	Lightcurves of the BL Lac object sample (6).	143
C.7	Lightcurves of the BL Lac object sample (7).	144
C.8	Lightcurves of the BL Lac object sample (8).	145
C.9	Lightcurves of the BL Lac object sample (9).	146
C.10	Lightcurves of the BL Lac object sample (10).	147
C.11	Lightcurves of the BL Lac object sample (11).	148

List of Tables

1.1	AGN taxonomy.	12
1.2	Size of AGN components.	14
1.3	Blazar evolution.	21
4.1	Observations of the Crab Nebula.	56
4.2	Fit parameters for the different Crab Nebula data samples.	58
5.1	List of MAGIC observation cycles.	59
5.2	Selection criteria	64
5.3	List of sources selected for the observation campaign	65
5.4	List of sources selected for the analysis	66
5.5	γ – hadron separation cut values.	72
5.6	Analysis results.	73
5.7	Crosscheck analysis results.	82
6.1	List of multi-wavelength data	87
6.2	Absorption coefficients and deabsorbed spectral information	88
6.3	Broad-band spectral indices $\alpha_{X-\gamma}$ of the BL Lac object sample.	92
6.4	Broad-band spectral indices $\alpha_{X-\gamma}$ for the low-state blazars.	96
A.1	Data sequences of the Crab Nebula (1).	101
A.2	Data sequences of the Crab Nebula (2).	102
A.3	Data sequences of the Crab Nebula (3).	102
A.4	Data sequences (1).	103
A.5	Data sequences (2).	104
A.6	Data sequences (3).	105
A.7	Data sequences (1).	106
A.8	Data sequences (5).	107

List of Tables

A.9 Data sequences (6). 108
A.10 Data sequences (7). 109
A.11 Data sequences (8). 110
A.12 Data sequences (9). 111
A.13 Data sequences (10). 112
A.14 Data sequences (11). 113
A.15 Data sequences (12). 114
A.16 Data sequences (13). 115
A.17 Data sequences (14). 116
A.18 Data sequences (15). 117
A.19 Data sequences (16). 118
A.20 Data sequences (17). 119
A.21 Data sequences of the crosscheck sample (1). 120
A.22 Data sequences of the crosscheck sample (2). 121
A.23 Data sequences of the crosscheck sample (3). 122

Bibliography

A. A. Abdo, M. Ackermann, M. Ajello, et al. Fermi Observations of TeV-Selected Active Galactic Nuclei. *ApJ*, 707:1310–1333, December 2009.

J. Abraham, P. Abreu, M. Aglietta, et al. Correlation of the Highest-Energy Cosmic Rays with Nearby Extragalactic Objects. *Science*, 318:938–943, November 2007.

J. Abraham, P. Abreu, M. Aglietta, et al. Correlation of the highest-energy cosmic rays with the positions of nearby active galactic nuclei. *Astroparticle Physics*, 29:188–204, April 2008.

V. Acciari, E. Aliu, T. Arlen, et al. Discovery of Very High Energy Gamma-ray Radiation from the BL Lac 1ES 0806+524. *ApJL*, 690:L126–L129, January 2009a.

V. A. Acciari, E. Aliu, T. Arlen, et al. A connection between star formation activity and cosmic rays in the starburst galaxy M 82. arXiv:0911.0873, November 2009b.

V. A. Acciari et al., A. A. Abdo, et al., S. D. Barber, and D. M. Terndrup. Discovery of very high energy gamma rays from PKS 1424+240 and multiwavelength constraints on its redshift. arXiv:0912.0730, December 2009.

F. Acero et al. Detection of Gamma Rays From a Starburst Galaxy. arXiv:0909.4651, September 2009.

F. Aharonian, A. Akhperjanian, M. Beilicke, et al. Is the giant radio galaxy M 87 a TeV gamma-ray emitter? *A&A*, 403:L1–L5, May 2003a.

F. Aharonian, A. Akhperjanian, M. Beilicke, et al. Observations of H1426+428 with HEGRA. Observations in 2002 and reanalysis of 1999 & 2000 data. *A&A*, 403:523–528, May 2003b.

F. Aharonian, A. Akhperjanian, M. Beilicke, et al. Observations of 54 Active Galactic Nuclei with the HEGRA system of Cherenkov telescopes. *A&A*, 421:529–537, July 2004.

Bibliography

F. Aharonian, A. G. Akhperjanian, G. Anton, et al. Discovery of Very High Energy γ-Ray Emission from Centaurus a with H.E.S.S. *ApJL*, 695:L40–L44, April 2009.

F. Aharonian, A. G. Akhperjanian, K.-M. Aye, et al. Observations of Mkn 421 in 2004 with HESS at large zenith angles. *A&A*, 437:95–99, July 2005.

F. Aharonian, A. G. Akhperjanian, U. Barres de Almeida, et al. New constraints on the mid-IR EBL from the HESS discovery of VHE γ-rays from 1ES 0229+200. *A&A*, 475:L9–L13, November 2007.

J. Albert, E. Aliu, H. Anderhub, et al. Discovery of Very High Energy Gamma Rays from 1ES 1218+30.4. *ApJL*, 642:L119–L122, May 2006.

J. Albert, E. Aliu, H. Anderhub, et al. Discovery of Very High Energy γ-Rays from 1ES 1011+496 at z = 0.212. *ApJL*, 667:L21–L24, September 2007a.

J. Albert, E. Aliu, H. Anderhub, et al. Observation of Very High Energy γ-Rays from the AGN 1ES 2344+514 in a Low Emission State with the MAGIC Telescope. *ApJ*, 662:892–899, June 2007b.

J. Albert, E. Aliu, H. Anderhub, et al. Observations of Markarian 421 with the MAGIC Telescope. *ApJ*, 663:125–138, July 2007c.

J. Albert, E. Aliu, H. Anderhub, et al. Variable Very High Energy γ-Ray Emission from Markarian 501. *ApJ*, 669:862–883, November 2007d.

J. Albert, E. Aliu, H. Anderhub, et al. Very high energy gamma-ray observations during moonlight and twilight with the MAGIC telescope. arXiv:astro-ph/0702475, February 2007e.

J. Albert, E. Aliu, H. Anderhub, et al. FADC signal reconstruction for the MAGIC telescope. *Nuclear Instruments and Methods in Physics Research A*, 594:407–419, September 2008a.

J. Albert, E. Aliu, H. Anderhub, et al. Probing quantum gravity using photons from a flare of the active galactic nucleus Markarian 501 observed by the MAGIC telescope. *Physics Letters B*, 668:253–257, October 2008b.

J. Albert, E. Aliu, H. Anderhub, et al. Systematic Search for VHE Gamma-Ray Emission from X-Ray-bright High-Frequency BL Lac Objects. *ApJ*, 681:944–953, July 2008c.

J. Albert, E. Aliu, H. Anderhub, et al. Upper Limit for γ-Ray Emission above 140 GeV from the Dwarf Spheroidal Galaxy Draco. *ApJ*, 679:428–431, May 2008d.

Bibliography

J. Albert, E. Aliu, H. Anderhub, et al. Very-High-Energy gamma rays from a Distant Quasar: How Transparent Is the Universe? *Science*, 320:1752–, June 2008e.

J. Albert, E. Aliu, H. Anderhub, et al. VHE γ-Ray Observation of the Crab Nebula and its Pulsar with the MAGIC Telescope. *ApJ*, 674:1037–1055, February 2008f.

J. Albert, E. Aliu, H. Anderhub, et al. MAGIC observations of PG 1553+113 during a multiwavelength campaign in July 2006. *A&A*, 493:467–469, January 2009.

E. Aliu, H. Anderhub, L. A. Antonelli, et al. Observation of Pulsed γ-Rays Above 25 GeV from the Crab Pulsar with MAGIC. *Science*, 322:1221–, November 2008.

E. Aliu, H. Anderhub, L. A. Antonelli, et al. Upper Limits on the VHE Gamma-Ray Emission from the Willman 1 Satellite Galaxy with the Magic Telescope. *ApJ*, 697:1299–1304, June 2009.

R. Aloisio, V. Berezinsky, P. Blasi, et al. A dip in the UHECR spectrum and the transition from galactic to extragalactic cosmic rays. *Astroparticle Physics*, 27:76–91, February 2007.

L. A. Anchordoqui, G. E. Romero, and J. A. Combi. Heavy nuclei at the end of the cosmic-ray spectrum? *Phys.Rev.D*, 60(10):103001, November 1999.

H. Anderhub, L. A. Antonelli, P. Antoranz, et al. Simultaneous Multiwavelength Observation of Mkn 501 in a Low State in 2006. *ApJ*, 705:1624–1631, November 2009.

J. R. P. Angel and H. S. Stockman. Optical and infrared polarization of active extragalactic objects. *ARA&A*, 18:321–361, 1980.

J. Arons. Magnetars in the Metagalaxy: An Origin for Ultra-High-Energy Cosmic Rays in the Nearby Universe. *ApJ*, 589:871–892, June 2003.

N. Bade, V. Beckmann, N. G. Douglas, et al. On the evolutionary behaviour of BL Lac objects. *A&A*, 334:459–472, June 1998.

H. M. Badran, T. C. Weeks, and The VERITAS Collaboration. TeV Gamma-Ray Observations from the blazar, 1ES2344 with the Whipple Cherenkov Imaging Telescope. In *International Cosmic Ray Conference*, volume 7 of *International Cosmic Ray Conference*, page 2653, August 2001.

L. Bassani, R. Landi, A. Malizia, et al. IGR J22517+2218=MG3 J225155+2217: A New Gamma-Ray Lighthouse in the Distant Universe. *ApJL*, 669:L1–L4, November 2007.

Bibliography

R. D. Blandford and R. L. Znajek. Electromagnetic extraction of energy from Kerr black holes. *MNRAS*, 179:433–456, May 1977.

E. Boldt and P. Ghosh. Cosmic rays from remnants of quasars? *MNRAS*, 307:491–494, August 1999.

M. Böttcher and C. D. Dermer. An Evolutionary Scenario for Blazar Unification. *ApJ*, 564:86–91, January 2002.

S. M. Bradbury, T. Deckers, D. Petry, et al. Detection of γ-rays above 1.5TeV from MKN 501. *A&A*, 320:L5–L8, April 1997.

T. Bretz and D. Dorner. MARS-The Cherenkov Observatory edition. In F. A. Aharonian, W. Hofmann, & F. Rieger, editor, *American Institute of Physics Conference Series*, volume 1085 of *American Institute of Physics Conference Series*, pages 664–667, December 2008.

A. Caccianiga and M. J. M. Marchã. The CLASS blazar survey: testing the blazar sequence. *MNRAS*, 348:937–954, March 2004.

J.-M. Casandjian and I. A. Grenier. A revised catalogue of EGRET γ-ray sources. *A&A*, 489:849–883, October 2008.

M. Catanese, C. W. Akerlof, H. M. Badran, et al. Discovery of Gamma-Ray Emission above 350 GeV from the BL Lacertae Object 1ES 2344+514. *ApJ*, 501:616, July 1998.

A. Cavaliere and V. D'Elia. The Blazar Main Sequence. *ApJ*, 571:226–233, May 2002.

A. Cavaliere and D. Malquori. The Evolution of the BL Lacertae Objects. *ApJL*, 516:L9–L12, May 1999.

J. J. Condon, W. D. Cotton, E. W. Greisen, et al. The NRAO VLA Sky Survey. *AJ*, 115:1693–1716, May 1998.

L. Costamante and G. Ghisellini. TeV candidate BL Lac objects. *A&A*, 384:56–71, March 2002.

V. D'Elia and A. Cavaliere. The Connection between BL Lacs and Flat-Spectrum Radio Quasars. In P. Padovani and C. M. Urry, editors, *Blazar Demographics and Physics*, volume 227 of *Astronomical Society of the Pacific Conference Series*, page 252, 2001.

C. D. Dermer and R. Schlickeiser. Model for the High-Energy Emission from Blazars. *ApJ*, 416:458, October 1993.

Bibliography

C. D. Dermer, R. Schlickeiser, and A. Mastichiadis. High-energy gamma radiation from extragalactic radio sources. *A&A*, 256:L27–L30, March 1992.

D. Donato, G. Ghisellini, G. Tagliaferri, and G. Fossati. Hard X-ray properties of blazars. *A&A*, 375:739–751, September 2001.

D. Dorner, K. Berger, T. Bretz, and M. Gaug. Data Management and Processing for the MAGIC Telescope. In *International Cosmic Ray Conference*, volume 5 of *International Cosmic Ray Conference*, page 175, 2005.

M. Doro. The Commissioning and Characterization of the Calibration System of the MAGIC Telescope. Master's thesis, Universita Degli Studi di Padova, 2004.

M. Elvis, D. Plummer, J. Schachter, and G. Fabbiano. The Einstein Slew Survey. *ApJS*, 80:257–303, May 1992.

R. Falomo and J. K. Kotilainen. Optical imaging of the host galaxies of X-ray selected BL Lacertae objects. *A&A*, 352:85–102, December 1999.

B. L. Fanaroff and J. M. Riley. The morphology of extragalactic radio sources of high and low luminosity. *MNRAS*, 167:31P–36P, May 1974.

G. Fossati, L. Maraschi, A. Celotti, et al. A unifying view of the spectral energy distributions of blazars. *MNRAS*, 299:433–448, September 1998.

M. Gaug. *Calibration of the MAGIC Telescope and Observation of Gamma Ray Burst*. PhD thesis, Universitat Autonoma de Barcelona, 2006.

G. Ghisellini, A. Celotti, and L. Costamante. Low power BL Lacertae objects and the blazar sequence. Clues on the particle acceleration process. *A&A*, 386:833–842, May 2002.

G. Ghisellini, A. Celotti, G. Fossati, et al. A theoretical unifying scheme for gamma-ray bright blazars. *MNRAS*, 301:451–468, December 1998.

G. Ghisellini and F. Tavecchio. The blazar sequence: a new perspective. *MNRAS*, 387:1669–1680, July 2008.

V. L. Ginzburg and S. I. Syrovatskii. Cosmic Magnetobremsstrahlung (synchrotron Radiation). *ARA&A*, 3:297, 1965.

Bibliography

P. Giommi, E. Massaro, P. Padovani, et al. ROXA J081009.9+384757.0: a 10^{47} erg s^{-1} blazar with hard X-ray synchrotron peak or a new type of radio loud AGN? *A&A*, 468:97–101, June 2007.

P. Giommi, M. T. Menna, and P. Padovani. The sedentary multifrequency survey - I. Statistical identification and cosmological properties of high-energy peaked BL Lacs. *MNRAS*, 310:465–475, December 1999.

P. Giommi, P. Padovani, M. Perri, et al. Parameter Correlations and Cosmological Properties of BL Lac Objects. In P. Giommi, E. Massaro, and G. Palumbo, editors, *Blazar Astrophysics with BeppoSAX and Other Observatories*, page 133, 2002a.

P. Giommi, M. Perri, S. Piranomonte, and P. Padovani. A New (NVSS-RASS) 1Jy Blazar Sample. In P. Giommi, E. Massaro, & G. Palumbo, editor, *Blazar Astrophysics with BeppoSAX and Other Observatories*, page 123, 2002b.

P. Giommi, S. Piranomonte, M. Perri, and P. Padovani. The Sedentary Survey of Extreme High Energy Peaked BL Lacs. II. The Catalog and Spectral Properties. arXiv:astro-ph/0411093, November 2004.

F. Goebel, H. Bartko, E. Carmona, et al. Upgrade of the MAGIC Telescope with a Multiplexed Fiber-Optic 2GSamples/s FADC Data Acquisition System system. In *International Cosmic Ray Conference*, volume 3 of *International Cosmic Ray Conference*, pages 1481–1484, 2008.

F. Goebel, K. Mase, M. Meyer, et al. Absolute energy scale calibration of the MAGIC telescope using muon images. In *International Cosmic Ray Conference*, volume 5 of *International Cosmic Ray Conference*, page 179, 2005.

R. C. Hartman, D. L. Bertsch, S. D. Bloom, et al. The Third EGRET Catalog of High-Energy Gamma-Ray Sources. *ApJS*, 123:79–202, July 1999.

D. Heck, J. Knapp, J. N. Capdevielle, et al. CORSIKA: A Monte Carlo code to simulate extensive air showers. *Wissenschaftliche Berichte, FZKA-6019. Forschungszentrum Karlsruhe*, 1998.

A. M. Hillas. The Origin of Ultra-High-Energy Cosmic Rays. *ARA&A*, 22:425–444, 1984.

A. M. Hillas. Cerenkov light images of EAS produced by primary gamma. In F. C. Jones, editor, *International Cosmic Ray Conference*, volume 3 of *International Cosmic Ray Conference*, pages 445–448, August 1985.

W. Hofmann and S. Fegan. H.E.S.S. and Fermi-LAT discovery of VHE and HE emission from blazar 1ES 0414+009. *The Astronomer's Telegram*, 2293:1, November 2009.

J. Holder, I. H. Bond, P. J. Boyle, et al. Detection of TeV Gamma Rays from the BL Lacertae Object 1ES 1959+650 with the Whipple 10 Meter Telescope. *ApJL*, 583:L9–L12, January 2003.

D. Horan, H. M. Badran, I. H. Bond, et al. Detection of the BL Lacertae Object H1426+428 at TeV Gamma-Ray Energies. *ApJ*, 571:753–762, June 2002.

Y. Inoue and T. Totani. The Blazar Sequence and the Cosmic Gamma-ray Background Radiation in the Fermi Era. *ApJ*, 702:523–536, September 2009.

Y. Inoue, T. Totani, S. Inoue, et al. The Cosmological Evolution of Blazars and the Extragalactic Gamma-Ray Background in the Fermi Era. arXiv:1001.0103, December 2010.

M. Irwin, S. Maddox, and R. McMahon. The APM Sky Catalogues. *IEEE Spectrum*, 2:14–16, June 1994.

C. Isola, G. Sigl, and G. Bertone. Ultra-High Energy Cosmic Rays from Quasar Remnants. arXiv:astro-ph/0312374, December 2003.

J. D. Jackson. *Klassische Elektrodynamik.* de Gruyter, 2nd edition, 1982.

T. M. Kneiske and H. Dole. A Lower-Limit Flux for the Extragalactic Background Light. arXiv:1001.2132, January 2010.

H. Kuehr, A. Witzel, I. I. K. Pauliny-Toth, and U. Nauber. A catalogue of extragalactic radio sources having flux densities greater than 1 Jy at 5 GHz. *A&AS*, 45:367–430, September 1981.

H. Landt, P. Padovani, E. S. Perlman, et al. The Deep X-Ray Radio Blazar Survey (DXRBS) - II. New identifications. *MNRAS*, 323:757–784, May 2001.

J. E. Ledden and S. L. Odell. The radio-optical-X-ray spectral flux distributions of blazars. *ApJ*, 298:630–643, November 1985.

R. W. Lessard, J. H. Buckley, V. Connaughton, and S. Le Bohec. A new analysis method for reconstructing the arrival direction of TeV gamma rays using a single imaging atmospheric Cherenkov telescope. *Astroparticle Physics*, 15:1–18, March 2001.

Bibliography

T.-P. Li and Y.-Q. Ma. Analysis methods for results in gamma-ray astronomy. *ApJ*, 272: 317–324, September 1983.

K. Mannheim. The proton blazar. *A&A*, 269:67–76, March 1993.

L. Maraschi, L. Foschini, G. Ghisellini, et al. Testing the blazar spectral sequence: X-ray-selected blazars. *MNRAS*, 391:1981–1993, December 2008a.

L. Maraschi, G. Ghisellini, and A. Celotti. A jet model for the gamma-ray emitting blazar 3C 279. *ApJL*, 397:L5–L9, September 1992.

L. Maraschi, G. Ghisellini, F. Tavecchio, et al. The Spectral Sequence of Blazars – Status and Perspectives. *IJMPD*, 17:1457–1466, 2008b.

M. J. Marchã, A. Caccianiga, I. W. A. Browne, and N. Jackson. The CLASS blazar survey - I. Selection criteria and radio properties. *MNRAS*, 326:1455–1466, October 2001.

A. P. Marscher, S. G. Jorstad, F. D. D'Arcangelo, et al. The inner jet of an active galactic nucleus as revealed by a radio-to-γ-ray outburst. *Nature*, 452:966–969, April 2008.

M. Meyer. *Observations of a systematically selected sample of high frequency peaked BL Lac objects with the MAGIC telescope*. PhD thesis, Bayerische Julius-Maximilians-Universität Würzburg, 2008.

A. Moralejo, M. Gaug, E. Carmona, and the MAGIC collaboration. MARS, the MAGIC Analysis and Reconstruction Software. arXiv:0907.0943, July 2009.

A. Mücke and R. J. Protheroe. A proton synchrotron blazar model for flaring in Markarian 501. *Astroparticle Physics*, 15:121–136, March 2001.

M. Nagano and A. A. Watson. Observations and implications of the ultrahigh-energy cosmic rays. *Reviews of Modern Physics*, 72:689–732, July 2000.

E. Nieppola, M. Tornikoski, and E. Valtaoja. Spectral energy distributions of a large sample of BL Lacertae objects. *A&A*, 445:441–450, January 2006.

K. Nilsson, M. Pasanen, L. O. Takalo, et al. Host galaxy subtraction of TeV candidate BL Lacertae objects. *A&A*, 475:199–207, November 2007.

K. Nilsson, T. Pursimo, J. Heidt, et al. R-band imaging of the host galaxies of RGB BL Lacertae objects. *A&A*, 400:95–118, March 2003.

Bibliography

R. A. Ong and P. Fortin. Discovery of High-Energy Gamma-Ray Emission from the BL Lac Object RBS 0413. *The Astronomer's Telegram*, 2272:1, October 2009.

P. Padovani, L. Costamante, G. Ghisellini, et al. BeppoSAX Observations of Synchrotron X-Ray Emission from Radio Quasars. *ApJ*, 581:895–911, December 2002.

P. Padovani, E. S. Perlman, H. Landt, et al. What Types of Jets Does Nature Make? A New Population of Radio Quasars. *ApJ*, 588:128–142, May 2003.

E. S. Perlman, P. Padovani, P. Giommi, et al. The Deep X-Ray Radio Blazar Survey. I. Methods and First Results. *AJ*, 115:1253–1294, April 1998.

D. Petry, S. M. Bradbury, A. Konopelko, et al. Detection of VHE γ-rays from MKN 421 with the HEGRA Cherenkov Telescopes. *A&A*, 311:L13–L16, July 1996.

M. Punch, C. W. Akerlof, M. F. Cawley, et al. Detection of TeV photons from the active galaxy Markarian 421. *Nature*, 358:477, August 1992.

J. Quinn, C. W. Akerlof, S. Biller, et al. Detection of Gamma Rays with E > 300 GeV from Markarian 501. *ApJL*, 456:L83+, January 1996.

M. J. Rees. Studies in radio source structure-III. Inverse Compton radiation from radio sources. *MNRAS*, 137:429, 1967.

W. A. Rolke, A. M. López, and J. Conrad. Limits and confidence intervals in the presence of nuisance parameters. *Nuclear Instruments and Methods in Physics Research A*, 551:493–503, October 2005.

J. H. Rose. Cherenkov Telescope Calibration using Muon Ring Images. In *International Cosmic Ray Conference*, volume 3 of *International Cosmic Ray Conference*, page 464, 1995.

S. Rügamer. Systematische Studien zur Verwendung der Zeitstruktur von Luftschauern zur Reduktion des Untergrundes in MAGIC-Daten. Master's thesis, Bayerische Julius-Maximilians-Universität Würzburg, 2006.

B. Sbarufatti, A. Treves, and R. Falomo. Imaging Redshifts of BL Lacertae Objects. *ApJ*, 635:173–179, December 2005.

B. Sbarufatti, A. Treves, R. Falomo, et al. ESO Very Large Telescope Optical Spectroscopy of BL Lacertae Objects. II. New Redshifts, Featureless Objects, and Classification Assessments. *AJ*, 132:1–19, July 2006.

Bibliography

R. Scarpa, C. M. Urry, R. Falomo, et al. The Hubble Space Telescope Survey of BL Lacertae Objects. I. Surface Brightness Profiles, Magnitudes, and Radii of Host Galaxies. *ApJ*, 532: 740–815, April 2000a.

R. Scarpa, C. M. Urry, P. Padovani, et al. The Hubble Space Telescope Survey of BL Lacertae Objects. IV. Infrared Imaging of Host Galaxies. *ApJ*, 544:258–268, November 2000b.

F. W. Stecker, O. C. de Jager, and M. H. Salamon. Predicted Extragalactic TeV Gamma-Ray Sources. *ApJL*, 473:L75+, December 1996.

M. Stickel, P. Padovani, C. M. Urry, et al. The complete sample of 1 Jansky BL Lacertae objects. I - Summary properties. *ApJ*, 374:431–439, June 1991.

A. W. Strong, I. V. Moskalenko, and O. Reimer. A New Determination of the Extragalactic Diffuse Gamma-Ray Background from EGRET Data. *ApJ*, 613:956–961, October 2004.

G. Tagliaferri, L. Foschini, G. Ghisellini, et al. Simultaneous Multiwavelength Observations of the Blazar 1ES 1959+650 at a Low TeV Flux. *ApJ*, 679:1029–1039, June 2008.

M. Teshima. MAGIC observes very high energy gamma ray emission from PKS 1424 +240. *The Astronomer's Telegram*, 2098:1, June 2009.

C. M. Urry and P. Padovani. Unified Schemes for Radio-Loud Active Galactic Nuclei. *PASP*, 107:803, September 1995.

M. Vietri. The Acceleration of Ultra–High-Energy Cosmic Rays in Gamma-Ray Bursts. *ApJ*, 453:883, November 1995.

W. Voges, B. Aschenbach, T. Boller, et al. The ROSAT all-sky survey bright source catalogue. *A&A*, 349:389–405, September 1999.

R. M. Wagner, M. Beilicke, F. Davies, H. Krawczynski, D. Mazin, M. Raue, S. Wagner, R. C. Walker, for the H. E. S. S. Collaboration, MAGIC Collaboration, VERITAS Collaboration, and the VLBA 43 GHz M87 monitoring team. A first joint M87 campaign in 2008 from radio to TeV gamma-rays. arXiv:0907.1465, July 2009a.

R. M. Wagner, E. J. Lindfors, A. Sillanpää, S. Wagner, and the CTA Consortium. The CTA Observatory. arXiv:0912.3742, December 2009b.

J. V. Wall and J. A. Peacock. Bright extragalactic radio sources at 2.7 GHz. III - The all-sky catalogue. *MNRAS*, 216:173–192, September 1985.

T. C. Weekes, M. F. Cawley, D. J. Fegan, et al. Observation of TeV gamma rays from the Crab nebula using the atmospheric Cerenkov imaging technique. *ApJ*, 342:379–395, July 1989.

T. Wibig and A. W. Wolfendale. At what particle energy do extragalactic cosmic rays start to predominate? arXiv:astro-ph/0410624, October 2004.

A. Wolter, I. M. Gioia, T. Maccacaro, et al. The number count distribution for X-ray-selected BL Lacertae objects and constraints on the luminosity function. *ApJ*, 369:314–319, March 1991.

www-01. http://heasarc.gsfc.nasa.gov/docs/history/, 2010.

www-02. http://www.pi1.physik.uni-erlangen.de/~kappes/lehre/WS05-VAT/V9/Hillas-plot.png, 2009.

www-03. F. Schmidt. CORSIKA shower images. http://www.ast.leeds.ac.uk/~fs/showerimages.html, 2009.

T. Yamamoto. The UHECR spectrum measured at the Pierre Auger Observatory and its astrophysical implications. In *International Cosmic Ray Conference*, volume 4 of *International Cosmic Ray Conference*, pages 335–338, 2008.

D. J. Yentis, R. G. Cruddace, H. Gursky, et al. The Cosmos/ukst Catalog of the Southern Sky. In H. T. MacGillivray & E. B. Thomson, editor, *Digitised Optical Sky Surveys*, volume 174 of *Astrophysics and Space Science Library*, page 67, 1992.

C. Zier and P. L. Biermann. Binary black holes and tori in AGN. II. Can stellar winds constitute a dusty torus? *A&A*, 396:91–108, December 2002.

Bibliography

List of Publications

Refereed Articles, proceedings and arXiv preprints

- Bretz, T, et al. (MAGIC Collaboration)
  "Comparison of on/off and wobble mode observations for MAGIC", 2005, Proc. 29th Int. Cosmic Ray Conf. (Pune), Vol. 4, 311-314

- Meyer, M., et al. (MAGIC Collaboration)
  "MAGIC observations of high-peaked BL Lacertae objects", 2005, Proc. 29th Int. Cosmic Ray Conf. (Pune), Vol. 4, 335-338

- Riegel, B., et al. (MAGIC Collaboration)
  "A systematic study of the interdependence of IACT image parameters", 2005, Proc. 29th Int. Cosmic Ray Conf. (Pune), Vol. 5, 215-218

- Albert, J., et al. (MAGIC Collaboration)
  "MAGIC Observations of Very High Energy γ-Rays from HESS J1813-178", 2006, ApJL, 637, L41

- Albert, J., et al. (MAGIC Collaboration)
  "Observation of Gamma Rays from the Galactic Center with the MAGIC Telescope", 2006, ApJL, 638, L101

- Albert, J., et al. (MAGIC Collaboration)
  "Observation of Very High Energy Gamma-Ray Emission from the Active Galactic Nucleus 1ES 1959+650 Using the MAGIC Telescope", 2006, ApJ, 639, 761

- Albert, J., et al. (MAGIC Collaboration)
  "Flux Upper Limit on Gamma-Ray Emission by GRB 050713a from MAGIC Telescope Observations", 2006, ApJL, 641, L9

List of publications

- Albert, J., et al. (MAGIC Collaboration)
  "Discovery of Very High Energy Gamma Rays from 1ES 1218+30.4", 2006, ApJL, 642, L119

- Albert, J., et al. (MAGIC Collaboration)
  "Observation of VHE Gamma Radiation from HESS J1834-087/W41 with the MAGIC Telescope", 2006, ApJL, 643, L53

- Albert, J., et al. (MAGIC Collaboration)
  "Variable Very-High-Energy Gamma-Ray Emission from the Microquasar LS I +61 303", 2006, Science, 312, 1771

- Albert, J., et al. (MAGIC Collaboration)
  "Discovery of Very High Energy γ-Rays from Markarian 180 Triggered by an Optical Outburst", 2006, ApJL, 648, L105

- Albert, J., et al. (MAGIC Collaboration)
  "Detection of Very High Energy Radiation from the BL Lacertae Object PG 1553+113 with the MAGIC Telescope", 2007, ApJL, 654, L119

- Albert, J., et al. (MAGIC Collaboration)
  "First Bounds on the Very High Energy γ-Ray Emission from Arp 220", 2007, ApJ, 658, 245

- Albert, J., et al. (MAGIC Collaboration)
  "Observation of Very High Energy γ-Rays from the AGN 1ES 2344+514 in a Low Emission State with the MAGIC Telescope", 2007, ApJ, 662, 892

- Meyer, M., et al. (MAGIC Collaboration)
  "Systematic search for VHE gamma-ray emission from X-ray bright high-frequency peaked BL Lac objects", 2007, Astronomische Nachrichten, 328, 621

- Albert, J., et al. (MAGIC Collaboration)
  "Observations of Markarian 421 with the MAGIC Telescope", 2007, ApJ, 663, 125

- Albert, J., et al. (MAGIC Collaboration)
  "Discovery of Very High Energy Gamma Radiation from IC 443 with the MAGIC Telescope", 2007, ApJL, 664, L87

List of publications

- Albert, J., et al. (MAGIC Collaboration)
  "Very High Energy Gamma-Ray Radiation from the Stellar Mass Black Hole Binary Cygnus X-1", 2007, ApJL, 665, L51

- Albert, J., et al. (MAGIC Collaboration)
  "Discovery of Very High Energy γ-Ray Emission from the Low-Frequency-peaked BL Lacertae Object BL Lacertae", 2007, ApJL, 666, L17

- Albert, J., et al. (MAGIC Collaboration)
  "MAGIC Upper Limits on the Very High Energy Emission from Gamma-Ray Bursts", 2007, ApJ, 667, 358

- Albert, J., et al. (MAGIC Collaboration)
  "Discovery of Very High Energy γ-Rays from 1ES 1011+496 at z = 0.212", 2007, ApJL, 667, L21

- Albert, J., et al. (MAGIC Collaboration)
  "The MAGIC Project: Contributions to ICRC 2007", 2007, arXiv:0709.3763

- Albert, J., et al. (MAGIC Collaboration)
  "Observation of VHE γ-rays from Cassiopeia A with the MAGIC telescope", 2007, A&A, 474, 937

- Albert, J., et al. (MAGIC Collaboration)
  "Variable Very High Energy γ-Ray Emission from Markarian 501", 2007, ApJ, 669, 862

- Albert, J., et al. (MAGIC Collaboration)
  "Constraints on the Steady and Pulsed Very High Energy Gamma-Ray Emission from Observations of PSR B1951+32/CTB 80 with the MAGIC Telescope", 2007, ApJ, 669, 1143

- Albert, J., et al. (MAGIC Collaboration)
  "VHE γ-Ray Observation of the Crab Nebula and its Pulsar with the MAGIC Telescope", 2008, ApJ, 674, 1037

- Albert, J., et al. (MAGIC Collaboration)
  "MAGIC Observations of the Unidentified γ-Ray Source TeV J2032+4130", 2008, ApJL, 675, L25

List of publications

- Albert, J., et al. (MAGIC Collaboration)
"Implementation of the Random Forest method for the Imaging Atmospheric Cherenkov Telescope MAGIC", 2008, Nuclear Instruments and Methods in Physics Research A, 588, 424

- Albert, J., et al. (MAGIC Collaboration)
"Upper Limit for γ-Ray Emission above 140 GeV from the Dwarf Spheroi-dal Galaxy Draco", 2008, ApJ, 679, 428

- Tagliaferri, G., et al. (i.a. MAGIC Collaboration)
"Simultaneous Multiwavelength Observations of the Blazar 1ES 1959+650 at a Low TeV Flux", 2008, ApJ, 679, 1029

- Albert, J., et al. (MAGIC Collaboration)
"Very-High-Energy gamma rays from a Distant Quasar: How Transparent Is the Universe?", 2008, Science, 320, 1752

- Albert, J., et al. (MAGIC Collaboration)
"Systematic Search for VHE Gamma-Ray Emission from X-Ray-bright High-Frequency BL Lac Objects", 2008, ApJ, 681, 944

- Albert, J., et al. (MAGIC Collaboration)
"Multiwavelength (Radio, X-Ray, and γ-Ray) Observations of the γ-Ray Binary LS I +61 303", 2008, ApJ, 684, 1351

- Albert, J., et al. (MAGIC Collaboration)
"Very High Energy Gamma-Ray Observations of Strong Flaring Activity in M87 in 2008 February", 2008, ApJL, 685, L23

- Aliu, E., et al. (MAGIC Collaboration)
"First Bounds on the High-Energy Emission from Isolated Wolf-Rayet Binary Systems", 2008, ApJL, 685, L71

- Albert, J., et al. (MAGIC Collaboration)
"FADC signal reconstruction for the MAGIC telescope", 2008, Nuclear Instruments and Methods in Physics Research A, 594, 407

- Albert, J., et al. (i.a. MAGIC Collaboration)
"Probing quantum gravity using photons from a flare of the active galactic nucleus Markarian 501 observed by the MAGIC telescope", 2008, Physics Letters B, 668, 253

List of publications

- Aliu, E., et al. (MAGIC Collaboration)
  "Observation of Pulsed γ-Rays Above 25 GeV from the Crab Pulsar with MAGIC", 2008, Science, 322, 1221

- Albert, J., et al. (MAGIC Collaboration)
  "MAGIC observations of PG 1553+113 during a multiwavelength campaign in July 2006", 2009, A&A, 493, 467

- Aliu, E., et al. (MAGIC Collaboration)
  "Improving the performance of the single-dish Cherenkov telescope MAGIC through the use of signal timing", 2009, Astroparticle Physics, 30, 293

- Donnarumma, I., et al. (i.a. MAGIC Collaboration)
  "The June 2008 Flare of Markarian 421 from Optical to TeV Energies", 2009, ApJL, 691, L13

- Aliu, E., et al. (MAGIC Collaboration)
  "Discovery of a Very High Energy Gamma-Ray Signal from the 3C 66A/B Region", 2009, ApJL, 692, L29

- Albert, J., et al. (MAGIC Collaboration)
  "Periodic Very High Energy γ-Ray Emission from LS I +61°303 Observed with the MAGIC Telescope", 2009, ApJ, 693, 303

- Anderhub, H., et al. (MAGIC Collaboration)
  "MAGIC upper limits to the VHE gamma-ray flux of 3C 454.3 in high emission state", 2009, A&A, 498, 83

- Aliu, E., et al. (MAGIC Collaboration)
  "Upper Limits on the VHE Gamma-Ray Emission from the Willman 1 Satellite Galaxy with the Magic Telescope", 2009, ApJ, 697, 1299

- Anderhub, H., et al. (MAGIC Collaboration)
  "MAGIC Collaboration: Contributions to the 31st International Cosmic Ray Conference (ICRC 2009)", 2009, arXiv:0907.0843

- Acciari, V. A., et al. (i.a. MAGIC Collaboration)
  "Radio Imaging of the Very-High-Energy γ-Ray Emission Region in the Central Engine of a Radio Galaxy", 2009, Science, 325, 444

List of publications

- Anderhub, H., et al. (MAGIC Collaboration)
  "Search for VHE γ-ray Emission from the Globular Cluster M13 with the Magic Telescope", 2009, ApJ, 702, 266

- Acciari, V. A., et al. (i.a. MAGIC Collaboration)
  "Simultaneous Multiwavelength Observations of Markarian 421 During Outburst", 2009, ApJ, 703, 169

- Anderhub, H., et al. (MAGIC Collaboration)
  "Discovery of very High Energy γ-Rays from the Blazar S5 0716+714", 2009, ApJL, 704, L129

- Seta, H., et al. (i.a. MAGIC Collaboration)
  "Suzaku and Multi-Wavelength Observations of OJ 287 during the Periodic Optical Outburst in 2007", 2009, PASJ, 61, 1011

- Anderhub, H., et al. (MAGIC Collaboration)
  "Simultaneous Multiwavelength Observation of Mkn 501 in a Low State in 2006", 2009, ApJ, 705, 1624

- Anderhub, H., et al. (MAGIC Collaboration)
  "Correlated X-Ray and Very High Energy Emission in the Gamma-Ray Binary LS I +61 303", 2009, ApJL, 706, L27

- Aleksić, J., et al. (MAGIC Collaboration)
  "Simultaneous multi-frequency observation of the unknown redshift bla-zar PG 1553+113 in March-April 2008", 2009, arXiv:0911.1088

- Aleksić, J., et al. (MAGIC Collaboration)
  "MAGIC TeV Gamma-Ray Observations of Markarian 421 during Multiwavelength Campaigns in 2006", 2010, arXiv:1001.1291

- Aleksić, J., et al. (MAGIC Collaboration)
  "MAGIC Gamma-ray Telescope Observation of the Perseus Cluster of Galaxies: Implications for Cosmic Rays, Dark Matter, and NGC 1275", 2010, ApJ, 710, 634

- Anderhub, H., et al. (MAGIC Collaboration)
  "Search for Very High Energy Gamma-ray Emission from Pulsar-Pulsar Wind Nebula Systems with the MAGIC Telescope", 2010, ApJ, 710, 828

List of publications

- Aleksić, J., et al. (MAGIC Collaboration, corresp. author: Daniel Höhne-Mönch)
  "Gamma-ray excess from a stacked sample of high-frequency peaked blazars observed with the MAGIC telescope", 2010, arXiv:1002.2951

- Aleksić, J., et al. (MAGIC Collaboration)
  "Search for an extended VHE gamma-ray emission from Mrk 421 and Mrk 501 with the MAGIC Telescope", 2010, arXiv:1004.1093

- Aleksić, J., et al. (MAGIC Collaboration)
  "MAGIC constraints on gamma-ray emission from Cygnus X-3", 2010, arXiv:1005.0740

Conference Contributions

- Höhne, D., et al. (MAGIC Collaboration)
  "Detection of VHE gamma-rays from the Bl Lac object 1ES2344+514 with the MAGIC telescope", talk at the "Frühjahrstagung der Deutschen Physikalischen Gesellschaft", Dortmund, 2006

- Höhne, D., et al. (MAGIC Collaboration)
  "Automation of the Monte Carlo production of IACTs", talk at the "Frühjahrstagung der Deutschen Physikalischen Gesellschaft", Freiburg, 2008

- Höhne-Mönch, D. (MAGIC Collaboration)
  "MAGIC Upper Limits on 13 X-ray bright high peaked BL Lac objects (HBLs)", talk at the "Frühjahrstagung der Deutschen Physikalischen Gesellschaft", München, 2009

List of publications

Acknowledgements

First and foremost I want to thank Prof. Dr. Karl Mannheim for giving me the possibility to work on this doctoral thesis. Without his input and useful discussions with him this thesis would not have been come to life.

The MAGIC Collaboration is gratefully acknowledged for the provision of proprietary data of the MAGIC telescope.

I also want to thank my colleagues: Thomas Bretz for providing the major part of the analysis Software MARS; Daniela Dorner for her never ending effort in improving and introducing to me the MAGIC database. For fruitful discussions I thank Markus Meyer, Karsten Berger, Dominik Elsässer and Sheetal Saxena and all with whom I had a good time at the Astronomy Chair. Last but not least my special thanks goes to Stefan Rügamer. He always had an open ear for my sometimes annoying and fussy problems.

Finally I thank my family for their unconditional support. They had a hard time with me during the writing of this thesis.

This work was funded by the BMBF and the DFG.

I want morebooks!

Buy your books fast and straightforward online - at one of world's fastest growing online book stores! Environmentally sound due to Print-on-Demand technologies.

Buy your books online at
www.morebooks.shop

Kaufen Sie Ihre Bücher schnell und unkompliziert online – auf einer der am schnellsten wachsenden Buchhandelsplattformen weltweit! Dank Print-On-Demand umwelt- und ressourcenschonend produziert.

Bücher schneller online kaufen
www.morebooks.shop

KS OmniScriptum Publishing
Brivibas gatve 197
LV-1039 Riga, Latvia
Telefax: +371 686 204 55

info@omniscriptum.com
www.omniscriptum.com

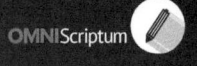

Printed by Books on Demand GmbH, Norderstedt / Germany